Advanced Thermoelectric Materials for Energy Harvesting Applications

Edited by Saim Memon

Published in London, United Kingdom

IntechOpen

Supporting open minds since 2005

Advanced Thermoelectric Materials for Energy Harvesting Applications
http://dx.doi.org/10.5772/intechopen.77430
Edited by Saim Memon

Contributors
Akira Suzuki, Shigeji Fujita, Dana Ben-Ayoun, Yaniv Gelbstein, Pablo Fredes, Ulrich Raff, Javiera Cuenca, Jose Pascal, Thiago Antonini Alves, Luis Vitorio Gulineli Fachini, Pedro Leineker Ochoski Machado, Larissa Krambeck, Romeu Miqueias Szmoski, Runar Unnthorsson, Robert Dell, Michael Thomas Petralia, Ashish Pokharel, Marcello Chiaberge, Dario Gandini, Lazaros Tzounis, Saim Memon

First published in London, United Kingdom, 2019 by IntechOpen
IntechOpen is the global imprint of INTECHOPEN LIMITED, registered in England and Wales, registration number: 11086078, 7th floor, 10 Lower Thames Street, London, EC3R 6AF, United Kingdom
Printed in Croatia

British Library Cataloguing-in-Publication Data
A catalogue record for this book is available from the British Library

Additional hard and PDF copies can be obtained from orders@intechopen.com

Advanced Thermoelectric Materials for Energy Harvesting Applications
Edited by Saim Memon
p. cm.
Print ISBN 978-1-78984-528-0
Online ISBN 978-1-78984-529-7
eBook (PDF) ISBN 978-1-83962-639-5

For EU product safety concerns:
IN TECH d.o.o., Prolaz Marije Krucifikse Kozulić 3, 51000 Rijeka, Croatia, info@intechopen.com or visit our website at intechopen.com.

Meet the editor

Dr Saim Memon is a senior lecturer in Electrical Engineering and Course Director HND-EEE at London South Bank University, London, UK. He holds a BEng (Hons) in Electrical Engineering, an MSc in Mechatronics, a PhD in Electrical and Electronic Engineering, and a PGCert Teaching Qualification - Further Education. He is a Chartered Engineer and Fellow of Higher Education Academy with Qualified Teacher Status (QTS) from the General Teaching Council for Scotland (GTCS). He teaches BEng/MEng/MSc modules and supervises BEng/MEng/MSc/PhD students. He has multi-disciplinary research/academic experience in electrical engineering, solar thermal vacuum engineering, and renewable energy engineering. His research interests include energy materials for vacuum-insulated smart windows, renewable energy technologies, and thermoelectric materials with vacuum insulation and heat storage. He has authored more than 35 publications and received numerous research grants and awards throughout his career. He has also developed research collaborations with leading scientists in the UK, Europe, USA, UAE, Japan, Thailand, Malaysia, Kenya, Peru, Pakistan, Russia, and China.

Contents

Preface

Electrical energy consumption is negatively affecting our environment and contributing to climate change. Therefore the research and industrial communities are working hard to minimize energy consumption using promising energy-efficient and renewable energy technologies. We know that it is possible to convert heat energy into electrical energy using thermoelectric devices; this heat energy can be from the sun or from an electro-mechanical device. However, thermoelectric devices traditionally suffer from lower efficiencies of energy conversion. This book, *Advanced Thermoelectric Materials for Energy Harvesting Applications*, is a research-intensive textbook consisting of eight chapters organized into three sections. Section 1 consists of Chapters 2, 3, and 4, which cover advanced thermoelectric materials and the topics of organic/inorganic thermoelectric materials, quantum theory of the Seebeck coefficient for the advancement of thermoelectric superconducting material, and the limits of Bismuth Telluride-based thermoelectric materials. Section 2, containing Chapters 5 and 6, evaluates behaviors and performance of thermoelectric devices. Section 3, containing Chapters 7 and 8, focuses on energy harvesting applications of thermoelectric devices. This book will be of interest to a wide range of individuals, such as scientists, engineers, researchers, and undergraduate and postgraduate students in the field of advanced thermoelectric materials.

I wish to thank all the authors for their contributions to this volume. I would like to express my appreciation to Ms. Ivana Barac of IntechOpen for her consistent support in publishing this book. And last but not least, I express great thanks to my family for their continuing support, motivation, and prayers.

Dr. Saim Memon PhD, MSc, BEng(Hons), PGC-TQFE, CEng, FHEA
Senior Lecturer in Electrical Engineering,
Course Director HND Electrical and Electronic Engineering,
Centre for Advanced Materials,
Division of Electrical and Electronic Engineering,
School of Engineering,
London South Bank University,
United Kingdom

Section 1

Advanced Thermoelectric Materials

Chapter 1

Introductory Chapter: Introduction to Advanced Thermoelectric Materials for Energy Harvesting Applications

Saim Memon

1. Introduction

Due to a cumulative trepidation of global carbon dioxide emissions and balancing the global electricity supply, an advancement into progressive technologies such as thermoelectrics has appeared to be promising [1]. With the fact that enormous amount of radiative heat from the sun, specifically in hot arid climate, and waste heat from electromechanical machines and devices can be harnessed. Over the past three decades, there has been an increasing significance of advancing the thermoelectric materials for energy harvesting applications [2] such as of passive cooling and utilizing the waste heat into useful electrical power, with Seebeck effect, from the automotive exhausts, combined heating and power devices, radiators [3], geothermal steam, electric vehicles and smart grids [4]. Thermoelectric generators have no mechanical moving parts but traditionally have lower efficiencies [5] and they are still in their development stage for the mass production scale for a wider industrial and domestic energy harvesting applications.

This book enlightens the design, implementation and thermoelectrical performance of existing and advanced thermoelectric materials for the vast number of energy harvesting applications. The authors lay the framework to emphasis from the quantum theory to the advanced development of the thermoelectric materials and bring attention to the questions of development, performance, evaluation and implementation of the advanced thermoelectric materials in the energy harvesting projects. To optimize the performance and apply thermoelectrics are of the paramount focus on this book. This book also discusses the methodologies and current state-of-the-art research in the field of thermoelectrics for energy conversion and management. The book establishes sustainable energy development goals of harnessing the waste heat energy from appliances, machines and devices and conversion to useful electrical power to a range of applications.

To appreciate the significance of advancing thermoelectric materials, here in particular the intention is to present the challenges that thermoelectric materials face today. However, thermoelectric material is a semiconducting material of two distinctive metals comprising p-type and n-type materials. These are accepted as the electron transport mechanism from the hot side to the cold side. Here, the Seebeck effect ascertains the characteristics of the materials, their series-parallel connections and of course thermally generated potential difference. So, the output-generated voltage with respect to the input (the temperature difference between two dissimilar metals) determines the efficiency and so far the maximum efficiency

IntechOpen

that traditionally reported was up to 5% with the use of Bismuth Telluride (Bi_2Te_3) composite. The limits of Bismuth Telluride (Bi_2Te_3)-based thermoelectric material and advancing it with the novel effects of doping the electron properties of Lead Telluride (PbTe) for the improvement of the thermoelectrical performances are discussed in this book. The recent methods for thermoelectric efficiency enhancement with $(PbTe)1\text{-}x(BiTe)x$ alloys could be used to improve electronic optimization.

There is a need of improving the efficiency of thermoelectric materials. Balancing the properties of the thermoelectric materials such as lowering the thermal conductivity, which is linked to figure of merit, is important. In this book, the organic (polymer and nanocomposite based) thermoelectric materials are discussed. The significance of organic and inorganic materials, technologies for the fabrication and performance parameters of thermoelectrics are discussed along with carbon-based materials such as graphene and carbon nanotubes for cost-effective solutions and enhancement of conversion efficiency. It is also important to understand the quantum theory of the Seebeck coefficient for the advancement of thermoelectric superconducting material which is also discussed in this book. This material incorporates the main superconducting properties such as zero resistance and superconducting temperature higher than the liquid nitrogen temperature.

Passive cooling using thermoelectric modules is usually applicable to remote locations where electricity from the grid is not possible, while utilizing the geothermal heat is of great importance in the future, for example, powering the closed-circuit television cameras, LED lightings and mobile cellular devices, which is discussed in detail in this book. A recent development in electronic mobile devices, lighting systems, concentrated quartz-halogen heat and batteries have allowed thermoelectric modules to use its generated power to add up the system efficiency. Thermoelectric generators are bridging their implication to electric vehicles, solar thermal and photovoltaic technologies, in which the scope of harnessing waste heat energy to electrical power is demanding with vacuum insulation [5, 6], and advancing the performance of thermoelectric materials [7] is of vital importance.

The question is what the optimum performance of using Bismuth Telluride (Bi_2Te_3)-based thermoelectric generator would be if the experimental setup is designed specifically for this? In this book, the experimental and simulated thermoelectrical performance parameters, such as the VI characteristic curves at variable temperatures, are analysed. In this, the focus is on harvesting the maximum possible waste direct heat to electrical power with the Bismuth Telluride (Bi_2Te_3)-based thermoelectric generator and demonstrating the experimental test setup necessary to achieve optimal performance for the recovery of the waste heat.

In terms of specific energy harvesting applications in the modern day, thermoelectric cooling devices could be used to operate ultraviolet LED in its nominal conditions in order to improve its optical performance. To harness the waste heat from the vehicles has already been an industrial automotive focus, and this book explains how waste heat from the exhaust system can be used to harness into useful electrical power.

However, this book does not stipulate the complete coverage of all the important advanced thermoelectric materials but specifically enlightens the submitted contributions by the authors. This book, nevertheless, provides state-of-the-art advanced thermoelectric materials and a range of energy harvesting technologies that might be helpful to readers to enhance their knowledge and develop eagerness to details in the science and development of thermoelectrics. It is intended that this book will motivate scientists, researchers, engineers and students to investigate further into the modern applications and materials development in the area of thermoelectrics.

Author details

Saim Memon
Centre for Advanced Materials, Division of Electrical and Electronic Engineering, School of Engineering, London South Bank University, London, UK

*Address all correspondence to: s.memon@lsbu.ac.uk

References

[1] Polozine A, Sirotinskaya S, Schaeffer L. History of development of thermoelectric materials for electric power generation and criteria of their quality. Materials Research. 2014;**17**(5):1260-1267

[2] Memon S, Tahir K. Experimental and analytical simulation analyses on the electrical performance of thermoelectric generator modules for direct and concentrated quartz-halogen heat harvesting. Energies. 2018;**11**(12):3315

[3] Nia MH, Nejad AA, Goudarzi AM, Valizadeh M, Samadian P. Cogeneration solar system using thermoelectric module and fresnel lens. Energy Conversion and Management. 2014;**84**:305-310

[4] Khan A, Memon S, Sattar TP. Analyzing integrated renewable energy and smart-grid systems to improve voltage quality and harmonic distortion losses at electric-vehicle charging stations. IEEE Access. 2018;**6**:26404-26415

[5] Kraemer D, Jie Q, McEnaney K, Cao F, Liu W, Weinstein LA, et al. Concentrating solar thermoelectric generators with a peak efficiency of 7.4%. Nature Energy. 2016;**1**(11):16153

[6] Memon S. Experimental measurement of hermetic edge seal's thermal conductivity for the thermal transmittance prediction of triple vacuum glazing. Case Studies in Thermal Engineering. 2017;**10**:169-178

[7] Ren GK, Wang S, Zhou Z, Li X, Yang J, Zhang W, et al. Complex electronic structure and compositing effect in high performance thermoelectric BiCuSeO. Nature Communications. 2019;**10**(1):2814

Chapter 2

Organic Thermoelectrics and Thermoelectric Generators (TEGs)

Lazaros Tzounis

Abstract

Inorganic and organic thermoelectric (TE) materials have received an extensive scientific interest during the last decades, due to their ability to directly convert the thermal energy to electricity. This is described by the well-known "Seebeck effect". TE materials can convert also electricity into cooling through the "Peltier effect". As such, TE materials and thermoelectric generator (TEG) devices can be utilized for potential applications including (i) thermal energy harvesting, (ii) local cooling and (iii) temperature sensing. The direct conversion of heat into electricity has been one of the most attractive solutions to the severe environmental and energy issues the humanity is coming across. This chapter covers the fundamental working principle of TE materials, the synthetic protocols for inorganic and organic thermoelectric materials, techniques and technologies for the fabrication of thermoelectric generators (otherwise defined as thermoelectric module devices) and a number of applications. Finally, future aspects and outlooks for further advancements at the "material" or "device" level for efficient power generation are remarked.

Keywords: Seebeck coefficient, thermoelectric (TE) nanomaterials, organic thermoelectrics, thermoelectric generators (TEGs)

1. Introduction to energy needs and wasted thermal energy

Due to the finite supply of fossil fuels and the human-induced global climate change, an emerging energy crisis has been realized in the twenty-first century giving rise to the exploitation of "green" energy and alternative energy resources [1–3]. There has been a substantial increase in the consumption of energy resources and especially that of petroleum feedstocks mainly due to (i) the industrial development and (ii) population growth [4]. In industrial environments and our daily life, large amounts of the generated heat energy cannot be effectively used getting inevitably wasted in the environment, e.g. emissions of factory boilers, car exhausts, friction, etc. A major contributor to waste heat is the transport sector, where only the 20% of the fuel's energy ends up as useful energy. Relatively, aeronautics and automotive are examples of high energy usage with low efficiency, where roughly 75% of the energy produced during combustion is lost in the turbine/exhaust or engine coolant in the form of heat. In relation, more than 60% of the energy produced in the USA is never utilized, as most of it is dissipated in the form

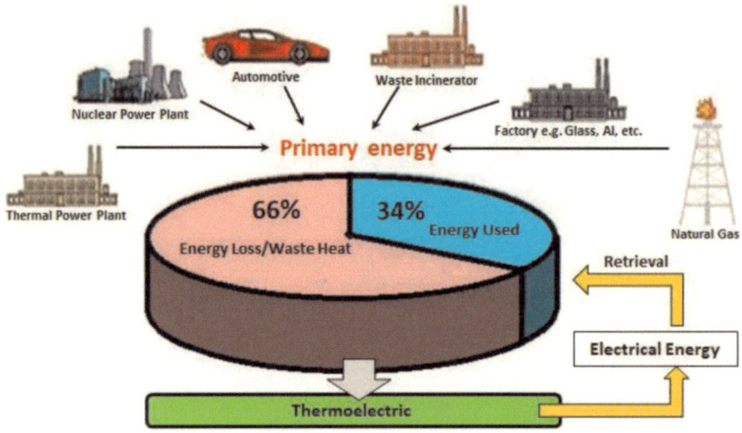

Figure 1.
Schematic representation of different sectors contributing to large amounts of wasted thermal energy.

of waste heat [5]. **Figure 1** shows the average effectively used energy from fossil fuels, while it represents schematically the different sectors that large amounts of waste heat are generated and could be transformed to electrical energy via the deployment of thermoelectric (TE) materials. Thus, a key solution for the huge amounts of wasted thermal energy (>60%) will be energy-harvesting technologies to effectively recycle and partially or fully harvest this inevitably generated and regrettably wasted thermal energy. Besides searching for alternative energy sources, for instance, solar energy, thermal energy, wind energy, hydrogen energy and biomass energy, to replace the conventional fossil fuels, improving the current efficiency of energy use is undoubtedly an expedient and viable solution.

2. Fundamentals of the thermoelectric effect and thermoelectric materials

The thermoelectric effect, otherwise known as "Seebeck effect", is the direct conversion of a temperature difference between two dissimilar electrical conductors or semiconductors to an electrical voltage. When the sides of TE materials are exposed to different temperatures, then a voltage is created across the two sides of the material. Conversely, when a voltage is applied, a temperature difference can be created, known as the "Peltier effect". At the atomic scale, when a temperature gradient is applied at the two end sides of a thermocouple, the electrons and holes move faster, and they have a lower density at the hot side, resulting in diffusion of electrons/holes towards the cold side as schematically demonstrated in **Figure 2**. This movement of carriers (electrons for n-type and holes for p-type materials) is translated into the generation of an electric field across the thermocouple. This is called as the "Seebeck effect", and the voltage created for a temperature difference, ΔT, under thermodynamic equilibrium is $S \times \Delta T$, where S is the Seebeck coefficient. TE materials are therefore one potential candidate for harvesting waste thermal energy, due to their ability to convert it into electricity, even under very-low-temperature gradients relative to the environmental temperature. This technology exhibits distinct advantages over other energy-harvesting technologies: (i) TE conversion is reliable and operates in silence as it works without mechanical movement, (ii) it is an environmentally friendly green technology, since no heat and no gaseous or chemical wastes are produced during operation, and (iii) it can be widely

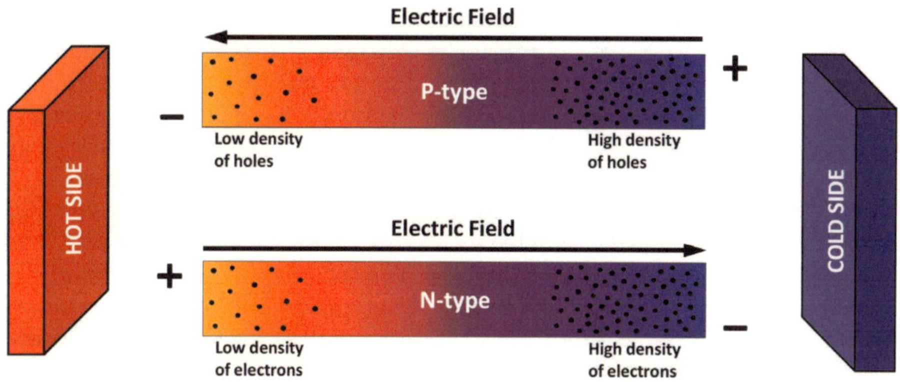

Figure 2.
A schematic representation of the TE effect showing the charge carriers of a p- or an n-type material to diffuse from the hot side to the cold, when a temperature gradient is applied.

used in places where other energy conversion technologies are unavailable, such as in the remote outer space, etc. [6, 7].

The "thermoelectric effect" encompasses three distinct effects: (a) the Seebeck effect, (b) the Peltier effect and (c) the Thomson effect. The Seebeck and Peltier effects are different manifestations of the same physical process, while it is often referred to as the Peltier-Seebeck effect. The Thomson effect is an extension of the Peltier-Seebeck model and is credited to Lord Kelvin. Seebeck in 1821 created a circuit made from two dissimilar metals; copper (Cu) and bismuth (Bi), while upon exposure the two end-junctions at different temperatures (hot/cold), he observed that the applied temperature gradient generated a voltage (experimentally observed that a compass magnet was deflected), which is commonly referred to then as the Seebeck effect [8]. Later, in 1834 Peltier observed that when an electric current flows through Cu and Bi wires at room temperature, a temperature difference could be created, and this is known onwards as the Peltier effect widely deployed in the refrigeration processes [9]. The Seebeck effect provides a theoretical basis for the applications of thermoelectric energy converters (energy harvesting), while the Peltier effect for cooling devices (refrigeration). To date, TE materials have been widely used in several high-tech applications such as aerospace, military, medical thermostats, microsensors, wearables, etc.

TE materials obey the thermoelectric or Seebeck effect described by the thermoelectric power (TEP), or thermopower, or Seebeck coefficient (S) [10]. The Seebeck coefficient is defined then as shown in Eq. (1):

$$S = \frac{\Delta V}{\Delta T} \tag{1}$$

where ΔV is the electric potential difference or the generated thermovoltage created by a temperature gradient, ΔT. It is an intrinsic material property related to the electronic properties, and it is positive for p-type and negative for n-type semiconductors [11]. The Seebeck coefficient is used for the calculation of the power factor ($PF = \sigma \times S^2$) (σ is the electrical conductivity), a well-known entity for comparing the voltage output (V_{out}) of different TE materials. The efficiency of TE materials is characterized by a dimensionless figure of merit (ZT); $ZT = (\sigma \times S^2/\kappa) \times T$, where σ represents the electric conductivity (S/m), S the Seebeck coefficient (μV/K), κ the material's thermal conductivity (W/mK) and T the absolute temperature (K) [12].

In recent years, great progress has been made in improving the ZT. It could be realized that an efficient TE material should exhibit high electrical conductivity combined with high Seebeck coefficient and low thermal conductivity. The Seebeck coefficient S, characteristic of the average entropy per charge transport, should be large in order to create a high voltage induced by a temperature gradient. The Seebeck coefficient, however, is not the only parameter to be optimized in order to maximize ZT. The electrical conductivity (σ) must be large to minimize the Joule heating during charge transport. Apart from the two parameters mentioned, a good TE material should exhibit also low thermal conductivity (κ), to prevent heat flow through the material. These three factors are interdependent in bulk TE materials, and altering one changes the other two. The difficulty in simultaneously optimizing them causes TE research to decay, until great reduction of thermal conductivity were both theoretically and experimentally proven in nanomaterials in 1993 [13]. The difficulty of designing high-performance TE materials arises from the fact that both electrical and thermal conductivities are related via the carrier concentration and, thus, optimizing one parameter will negatively affect the other. This interdependence has delayed the development of TE materials for many years.

Although TE technology possesses many merits and has been known for two centuries, it has only been applied in narrow fields because of its low conversion efficiency (typically less than 6%) [14]. The conversion efficiency strongly depends on the ZT of TE materials, as described by Eq. (2):

$$n = n_c \frac{\sqrt{1+ZT}-1}{\sqrt{1+ZT}+\frac{T_c}{T_h}} \tag{2}$$

where n is the conversion efficiency of heat to electricity; T_c and T_h are the temperatures of cold and hot sides of a TE material, respectively; and n_c represents the Carnot efficiency expressed as $n_c = 1 - T_c/T_h$ or $(T_h - T_c)/T_h \times 100\%$. It can be clearly realized that ZT should be at least above 3 in order to ensure that the TE material/device conversion efficiency is competitive with that of traditional power generators, which can reach 40% of Carnot efficiency [15]. The electrical conductivity (σ) being an important factor in the ZT equation can be expressed by Eq. (3):

$$\sigma = n\, e\, \mu \tag{3}$$

where n is the charge carrier density or concentration, μ is the mobility of charge carriers and e is the charge of unit carrier (electron or hole).

The electrical conductivity can be improved via chemical doping that each dopant atom can have one more valence electron than its host atoms and can therefore facilitate the increase of charge carrier density. Meanwhile, the dopants can reduce the mobility of charge carriers due to the enhanced scattering between dopants and carriers. The ideal density of charge carriers has been reported to be in the range of 10^{19} to 10^{21} cm^{-3} [16].

Doping improves also the Seebeck coefficient by changing the electron density of states (DOS) [17]. As shown in Eq. (4), the Seebeck coefficient is mainly affected by the charge carrier concentration, as well as the effective mass of the charge carriers (m^*), which usually decreases with increasing carrier mobility (k_B and h are the Boltzmann constant and Planck constant, respectively):

$$S = \frac{8\,\pi^2 k_B^2}{3\,e\;h^2} m^* T \left(\frac{\pi}{3n}\right)^{\frac{2}{3}} \tag{4}$$

The thermal conductivity is decreased by phonon scattering (e.g. phonon-boundary scattering, phonon-defect scattering and phonon-phonon scattering).

The thermal conductivity is the summation of two components: (i) the electron thermal conductivity (k_e) and (ii) the lattice thermal conductivity (k_c), as can be expressed in Eq. (5):

$$k = k_e + k_c \tag{5}$$

k_e is proportional to the electrical conductivity according to the Wiedemann-Franz law [18]. In semiconductors, usually >90% of thermal conductivity arises from the lattice thermal conductivity (k_c), which is independent of the electrical conductivity. Hence, reducing the lattice thermal conductivity will lead to a pronounced enhancement of the TE performance.

Then, the dimensionless "figure of merit", ZT, could be rewritten as shown in Eq. (6):

$$ZT = \left\{ \left[\frac{8}{3}\frac{\pi^2 k_B^2}{q\ h^2} m^* T \left(\frac{\pi}{3n}\right)^{\frac{2}{3}} \right]^2 \right\} \left(\frac{\mu}{k}\right) qT \tag{6}$$

where S, σ, k and T are the Seebeck coefficient, electrical conductivity, thermal conductivity and absolute temperature, respectively; k_B is the Boltzmann constant; h is Planck's constant; q is the charge; n is the charge carrier concentration; μ is the mobility and m^* is the effective mass of the charge carrier.

From the above, it could be realized that the Seebeck coefficient, thermal conductivity and electrical conductivity are strongly dependent, and it is therefore very challenging to improve them simultaneously in bulk TE materials. This is mainly the reason that very few researches exist on TE materials from the 1960s to the 1990s, until the article was published in 1993 by Dresselhaus and co-workers showing theoretically that low-dimensional materials can have higher ZT than their bulk analogues, due to both their lower thermal conductivity and quantum confinement effects [19, 20]. This work triggered the scientific community research interest in TE materials and provided a mainstream approach and strategy for the enhancement of TE material's performance and ZT values. This can be achieved more precisely via creating proper material "nanostructuring" inducing a large number of interfaces and facilitating thus a phonon scattering resulting into low thermal conductivity values. The TE material's research and the promising nanostructuration approaches are part of the wider field of nanotechnology, where key elements are the long-range ordering with controlled nanostructures for enhanced optical [21–29], electrical [30–39], mechanical [40–42] and magnetic [43, 44] properties.

Figure 3 demonstrates the number of publications on TE materials from 1965 to today, highlighting the slight increase of TE research rate from 1965 to 1993 and the rapid increase from that year to now, due to the flagship stimulating work of Dresselhaus (data collected from Science Direct data library using "thermoelectric" as the keyword). The extreme interest for TE materials during the last decades, mainly during the last decade, has been the driving force for many review articles that have appeared focusing on the different types of TE materials [45–51] including oxide [45] and organic based [47]. For example, Sootsman et al. summarized new and old concepts in inorganic TE materials in 2009 [48]. M. S. Dresselhaus and co-workers had reviewed in 2007 new directions for low-dimensional inorganic TE materials [13]. Li et al. highlighted progress in TE materials with high ZT and the related fabrication processes for producing nanostructured materials including Bi-Te alloys, skutterudite compounds, Ag-Pb-Sb-Te quaternary systems, half-Heusler compounds and high ZT oxides [52]. The last decade, reviews on organic-based TE materials and devices have appeared. Relatively, Du et al. reviewed in

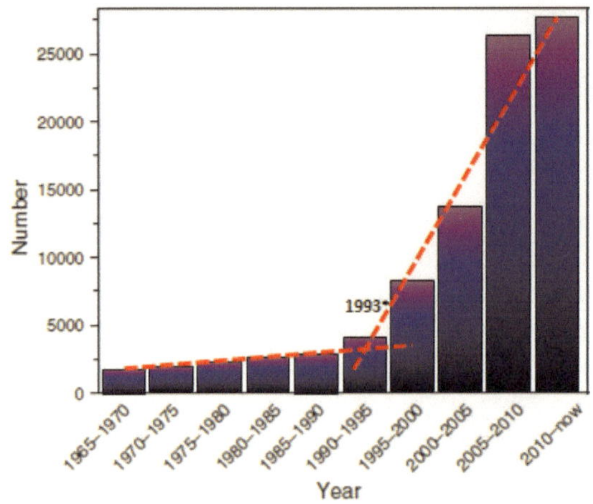

Figure 3.
The number of publications in the area of thermoelectric materials and related technology from 1965 to today. The data were collected from science direct data library using "thermoelectric" as the keyword.

2012 the research progress on polymer-inorganic TE nanocomposite materials [53]. Gao et al. summarized in 2016 the conducting polymer/carbon particle TE nanocomposites as emerging green energy materials [54]. Chen et al. reviewed in 2017 advances in polymer TE composites [55]. A very important review by He et al. in 2017 had summarized all the state-of-the-art TE devices, architectures, geometries, device interconnections, contact optimization, etc., based on inorganic and organic (small molecule, polymer, micro- and nanocomposites) TE materials [56]. Finally, it is worth mentioning that, currently, the highest value of ZT (ZT = 3.6 at 580 K) has been reported by Harman et al. for $PbSe_{0.98}Te_{0.02}/PbTe$ quantum dot superlattices (QDSLs) grown by molecular beam epitaxy (MBE) [57].

3. Organic thermoelectrics: polymer and nanocomposite systems

3.1 Conjugated polymer thermoelectric materials

Organic thermoelectric (OTE) materials or "organic thermoelectrics" have attracted an increased scientific interest as an alternative approach to conventional inorganic thermoelectrics. Conducting polymers have been suggested as TE materials for potential large-area thermoelectric applications. This is because OTEs are compatible with inexpensive, large-scale processing methods and often possess unique mechanical flexibility, which makes them geometrically versatile to be integrated in complex three-dimensional (3D)-shaped objects, e.g. wearables, car roofs, etc.

Till now, polyaniline (PANI) has been the most studied conductive polymer due to its high electrical conductivity, good chemical and thermal stability and ease of preparation and processing/casting from solution. Poly(3,4-ethylenedioxythiophene):polystyrenesulfonate (PEDOT:PSS) is the second most used polymer with potential processability from water dispersions that are commercially available, good charge transport properties (both of electronic carriers, as well as ions) resulting in relatively high electrical conductivities. However, PEDOT:PSS exhibits low resistance to humid environments due to its hygroscopic behaviour arising from the PSS phase.

In 2002, the first study appeared in literature reporting on the use of electrically conductive polymers and especially polyaniline films as organic thermoelectric

(OTE) materials [58]. OTEs are particularly attractive for low-quality waste heat harvesting such as waste heat at low temperatures. Till then, high *ZT* values (>0.1) have been achieved for both p-type and n-type OTE materials [59, 60]. Similarly to inorganic thermoelectrics, organic ones should be also optimally "electronic crystals and phonon glasses". The main crucial parameter to tune and optimize for organic thermoelectrics is to achieve "electronic crystals", since their thermal conductivities are intrinsically relatively low. Therefore, the thermoelectric parameter to put effort on and optimize is the power factor, via simultaneously increasing the electrical conductivity and Seebeck coefficient. The Seebeck coefficient is directly related to the DOS as stated several times in this chapter, reflecting the average entropy transported per charge carrier. As such, it decreases with increasing carrier concentration since the mobility is decreased. On the contrary, the electrical conductivity increases with carrier concentration, n, and carrier mobility μ ($\sigma = ne\mu$).

Organic semiconductors have been largely neglected as TE materials, despite their inherent low thermal conductivities (≈0.3 W/mK) and high electrical conductivities (>1000 S/cm) [47]. This is due to the fact that conducting polymers are not stable at high temperatures with their maximum operational temperatures limited in the range of 200–250°C. Hence, the benchmark for polymer-based thermoelectrics is Bi_2Te_3 alloys that exhibit a *ZT* of 1.2 at room temperature [61].

Conjugated polymers and mainly PANI [62–64] and polythiophenes [65–69] have been investigated as OTE materials. **Figure 4** depicts the molecular structures of representative p- and n-type semiconducting polymers and dopants, associated with the power factor values that have been experimentally determined.

The most used p-type polymeric material investigated till now is the blend of poly(3,4-ethylenedioxythiophene):polystyrenesulfonate which has been utilized in various organic electronic devices, e.g. OPVs, OTFTs, OTEs, OLEDs, etc. PEDOT:PSS is a polyelectrolyte consisting of positively charged PEDOT and negatively charged PSS. Crispin et al. reported power factors of 300 μW m^{-1} K^{-2} and *ZT* values of 0.25, after de-doping highly conductive PEDOT:tosylate with tetrakis (dimethylamino)ethylene [47]. Besides, the addition of carbon nanotubes (CNTs) or graphene is a popular approach to modulate the TE properties of PEDOT:PSS

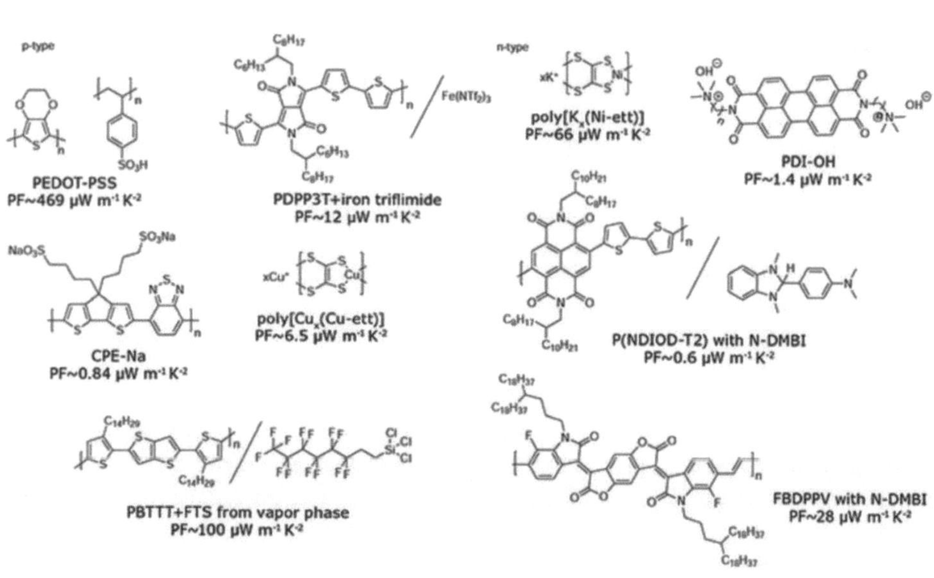

Figure 4.
Representative p-type and n-type semiconducting polymers and dopant materials together with the respective power factors.

[70]. By carefully structuring the composite film, conductivities of 105 S/m and Seebeck coefficients of 120 μV/K were measured, leading to power factors of 2710 μW m^{-1} K^{-2}, one of the highest values reported for OTE materials.

While the p-doping of organic semiconductors can be readily achieved, n-doping is more challenging. Organic electron-deficient semiconductors are associated with the high electron affinities (−3 to −4 eV), making the negatively charged molecules prone to reactions with environmental moisture or oxygen [71]. The charge transfer co-crystal salt of tetrathiafulvalene (TTF) and tetracyanoquinodimethane (TCNQ) is probably the most studied charge transfer salt showing promising n-type thermoelectric properties. Electrical conductivities of 500 S/cm and power factors of up to 40 μW m^{-1} K^{-2} have been reported. However, several drawbacks significantly limit the applicability of charge transfer crystals as TE materials, e.g. modulating the carrier densities is difficult since the stoichiometry of the co-crystals must be accurately respected and the physical properties of the co-crystals are not isotropic. Alternative n-type conductors for TE applications have mainly focused on the perylenediimide- and naphtalenediimide-containing organic semiconductors. Segalman et al. synthesized a series of perylenediimide (PDI)-based molecular semiconductors functionalized with tertiary amine-containing side chains [72]. Upon thermal annealing, the functionalized PDI moieties self-dope via a dehydration reaction of the tethered tertiary ammonium hydroxide. By carefully designing the side chains, the self-doped PDI moieties achieve conductivities of 0.5 S/cm and power factors of 1.4 μW m^{-1} K^{-2}. Chabinyc et al. extrinsically doped the high-performing n-type polymer poly(N,N′-bis(2-octyldodecyl)-1,4,5,8-napthalenedicarboximide-2,6-diyl]-alt-5,5′-(2,2′-bithiophene)) (P(NDIOD-T2) with the molecular dopant (4-(1,3-dimethyl-2,3-dihydro-1H-benzoimidazol-2-yl) phenyl) (N-DMBI) [73]. While the conductivity initially increases as a function of dopant loading, a sharp drop in conductivity was observed at higher loadings. The miscibility of the N-DMBI dopant in the polymer phase is limited, which is why at higher dopant loadings, the dopant crystallizes and phase separates from the polymer matrix, thus reducing the doping efficiency. Despite the morphological instabilities, Seebeck coefficients of −850 μV/K have been achieved with power factors of 0.6 μW m^{-1} K^{-2}. In a recent report, Pei et al. showed that BDOPV-based FBDPPV polymers have reached a record power factor of 28 μW m^{-1} K^{-2} [74]. Huang et al. reported that thiophene-diketopyrrolopyrrole-based quinoidal (TDPPQ) can exhibit a high power factor of 113 μW m^{-1} K^{-2}, when the material is interfacially doped by the bismuth. The performance is the best value for all reported n-type small molecules [75].

3.2 Nanocomposite polymer thermoelectric materials

Nanocomposite polymer thermoelectrics using a conducting and nonconducting matrix and organic (e.g. CNTs, graphene oxide, fullerenes, etc.) or inorganic nanoinclusions (e.g. Bi_2Te_3, PbTe, Te nanowires, etc.) have been also extensively studied.

PEDOT:PSS (CLEVIOS PH1000) mixed with Bi_2Te_3 particles have reached power factors ($PF = \sigma \times S^2$, where σ is the electric conductivity and S the Seebeck coefficient) in the range of ~130 μW/m K^2 [76]. The incorporation of CNTs may enhance their performance via increased conductivity or molecular orientation effects of the polymer chains. Thereby, high filler loadings (>> 50 wt.%) can be realized [77], resulting in high electrical conductivities [78–81]. Namely, electrical conductivities up to 4×10^5 S/m and power factors in the range of ~140 μW m^{-1} K^{-2} have been reported by Moriarty et al. for single-walled carbon nanotubes (SWCNTs) in a PEDOT:PSS matrix [68]. However, the low thermal and moisture stability of these materials is an impediment to engineering and structural applications.

CNTs have been introduced also via solution and melt-mixing methods in engineering nonconductive thermoplastic polymer matrices resulting into nanocomposites with thermal energy-harvesting property [82–86]. Polymer nanocomposites are attractive organic materials due to their ease of production, relatively low cost, flexibility and high specific properties [30–33, 87, 88]. SWCNT polycarbonates (PC)/SWCNT nanocomposites prepared by solvent mixing showed that by increasing the SWCNT content (up to 30 wt.%), the electrical conductivity increases to approximately 1000 S/m, and the Seebeck coefficient reaches 60 μV/K with only a slight dependence on the SWCNT content. CNT composites with polymers having electron-rich functional groups, like PVA and polyethyleneimine (PEI), have been found to act as n-doping to the incorporated SWCNTs and resulted in coefficients up to −21.5 μV/K. Antar et al. reported on melt-mixed composites of polylactide (PLA) with multi-walled carbon nanotubes (MWCNTs) and expanded graphite with high filling levels (up to 30 wt.%) resulting in electrical conductivities of ~4000 S/m [89]. The Seebeck coefficient reached a maximum of 17 μV/K for the composites with expanded graphite and ~9 μV/K for MWCNT ones. Research from our group using a series of melt-mixed polycarbonate-MWCNT nanocomposites has shown that an increasing filler content results in an increase of the power factor due to the increase of the electrical conductivity [36, 90, 91]. The TE properties of melt-mixed conductive nanocomposites of polypropylene (PP) filled with single-walled carbon nanotubes (2 wt.%) and copper oxide (5 wt.%) showed that by adding polyethylene glycol (PEG) during melt mixing p-type composites switched into n-type with Seebeck coefficient up to +45 μV/K and −56 μV/K, respectively [86].

Hierarchical CNT-coated fibrous reinforcement structures have been reported also as TE reinforcements upon their incorporation in polymer matrices for potential large-scale thermal energy harvesting by structural composites, e.g. in aerospace and automotive [34, 42]. Most of the polymeric matrices studied so far are based on aliphatic or semi-aromatic backbones. This severely limits their applicability as engineering materials capable of operating in high-temperature environments.

High-performance engineering polymers such as all-aromatic polyimides and poly(ether-imide)s (PEIs) are capable of withstanding high temperatures (>200°C) and exhibit glass-transition temperatures (T_g) above 200°C with superlative mechanical properties. Recently, Tzounis et al. [38] demonstrated the first synthesis of all-aromatic PEI-SWCNT nanocomposite films as thermoelectrics.

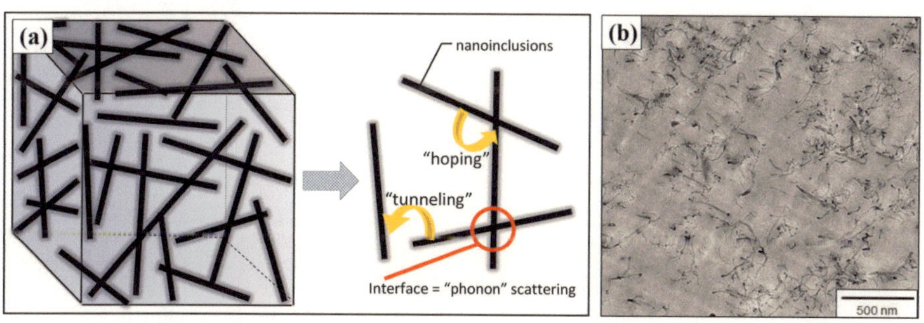

Figure 5.
(a) Schematic representation of the mechanism based on electronic charge transport that governs the TE effect upon exposure to a temperature gradient and (b) a TEM image of a polymer/CNT nanocomposite.

Polymer	Nanofiller	Preparation method	σ (S/cm)	S (μV/K)	PF (μW $m^{-1} K^{-2}$)	κ (W $m^{-1} K^{-1}$)	ZT (T)	Ref.
Polyacetylene	—	Casting	4.990–11.560	11.4–28.4	—	0.7	0.0047–0.38 (300 K)	[101]
Polypyrrole	—	Casting	100	12	—	0.1–0.2	0.002 (300 K)	[101]
Polypyrrole	rGO (67 wt.%)	In situ chemical polymerization	41.6	26.9	3.01 (300 K)	—	—	[102]
Polyaniline (PANI)	—	Casting	7.000	7	—	0.1–0.2	0.051 (300 K)	[101]
Polyaniline (PANI)	GNs (30 wt.%), HCl	In situ chemical polymerization	38	26	2.6 (453 K)	0.6	1.95×10^{-3} (453 K)	[103]
Polyaniline (PANI)	Bi_2Te_3	Solution mixing	60	110	5.1 (350 K)	—	0.18 (350 K) estimated	[104]
Polythiophene	—	Casting	100	21	—	0.1–0.2	0.0066 (300 K)	[101]
Poly(para-phenylene)	—	Casting	10^{-5}	12	—	0.1–0.2	2.1×10^{-10} (300 K)	[101]
Poly(p-phenylene vinylene)	—	Casting	10^{-5}	7	—	0,1–0,2	$7{,}2 \times 10^{-11}$ (300 K)	[101]
Poly(carbazolenevinylene)	—	Casting	5×10^{-3}	230	—	0.1–0.2	8×10^{-5} (300 K)	[101]
PEDOT:PSS	—	Casting	55	13	—	0.1–0.2	0.0014 (300 K)	[101]
PEDOT:PSS	—	Casting	900	75	—	0.24	0.42 (300 K)	[101]
PEDOT:PSS	rGO (21 wt.%)	Mixing	715.03	22.9	32.4 (300 K)	~0.2	0.067 (300 K)	[105]
PEDOT:PSS	Graphene (2 wt.%)	Solution spin casting	32.13	58.77	11.09 (300 K)	0.14	0.021 (300 K)	[106]
PEDOT:PSS	Fullerene (9%) rGO (16%)	In situ chemical polymerization	50.8 ± 5.9	31.8 ± 2.3	5.2 ± 0.9 (300 K)	—	—	[107]
PEDOT:PSS	SWCNT (DMSO)	Film casting	~4000 (95 wt.%)	14–26	~140 (85 wt.%)	0.4–0.7	0.03 (40 wt.%) (300 K)	[108]
PEDOT:PSS	Tos	Film casting	6×10^{-4}–300	40–780	—	0.37	0–0.25 (300 K)	[109]
PEDOT:PSS	Te	Films	19.3 ± 2.3	163 ± 4	70.9	0.22–0.30	0.10 (300 K)	[110]

Table 1.
A summary of the TE properties for eight representative conjugated polymer-based thermoelectrics [101].

Semicrystalline nanocomposites of PEI/SWCNT (10 vol.%) reached a maximum power factor of ~1.8 μW m^{-1} K^{-2}.

In a polymer (insulating or conjugated) matrix, nanoinclusions have the ability to allow electron (n-doping) or hole (p-doping) transport by a plausible tunneling or hopping mechanism, while at the same time, phonon scattering occurs at the nanoparticle-polymer-nanoparticle interfaces preventing their effective transmission and resulting in low thermal conductivities. Therefore, nanocomposites are considered as promising TE materials for thermoelectric power generation (used for the fabrication of TEGs), and a continuous demand remains for the increase of their electrical conductivity, Seebeck coefficient and power factor values compared to existing reported values.

Figure 5a shows the mechanism that governs the TE mechanism of a polymer nanocomposite, and **Figure 5b** depicts a representative transmission electron microscopy (TEM) image of a polymer/CNT TE nanocomposite showing the dispersed 1D nanoparticulates within the polymer matrix (from the author's personal microscopy investigation result data).

Besides, carbon nanomaterials as thin films and thick self-standing films (buckypapers) have been reported also as thermoelectrics. CNTs, for instance, are well-known for their semiconductor electronic properties and have shown thus promising TE performance. Their TE efficiency has been found to be enhanced by the level of doping [92, 93], as well as the dopant nature [34, 36, 39, 90, 94–97]. Doping of the SWCNTs by using, e.g. polyethyleneimine or hexafluoroacetone [98, 99], resulted in Seebeck coefficients up to −50 μV/K. Hewitt et al. reported Seebeck coefficients of CNT buckypapers between 11 and 19 μV/K and also discussed the dependence of the Seebeck coefficient on the CNT acidic treatment protocol [92]. Recently, Dörling et al. demonstrated that nitrogen-doping of the CNT graphitic lattice results in n-type TE behaviour [100].

Table 1 summarizes the electrical conductivity, Seebeck coefficient, thermal conductivity, power factor and *ZT* of different representative conjugated polymer-based TE materials with or without nanoinclusions.

4. Working principle and specific architectures of thermoelectric generators

Thermoelectric generators (TEGs) convert heat into electricity through the Seebeck effect. An applied temperature gradient across the generator will force heat to flow from the hot to the cold side by thermal conduction, while some of this heat is converted to electricity. The possibility of converting a heat flux into an electrical current is realized by a TEG device (schematically illustrated in **Figure 6a**), while the opposite phenomenon is realised by a thermoelectric cooler (TEC) or Peltier device (**Figure 6b**). Regarding the Seebeck effect, when the junctions at the top are heated and those at the bottom are cooled, a temperature difference will occur. The electron/hole pairs are created at the hot end by absorbing heat and then recombined and liberated heat at the cold end. Driven by the mobility of hole/electron, the Seebeck voltage generates between the two ends, resulting in a current flow. As for the process of TE cooling or "Peltier effect", when a voltage is applied across a p/n junction, electron/hole pairs are generated in the vicinity of the junction and flow away, leading to the cooling of the junction on one end and the heating on the other end.

For an ideal TE device with constant TE properties, the maximum heat to electrical power conversion efficiency (η_{max}) and the output power density (P_{max}) are expressed as

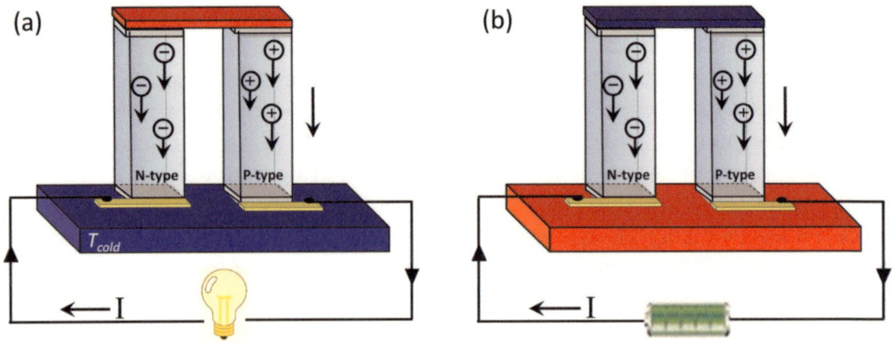

Figure 6.
Operation principle of a (a) thermoelectric generator (TEG) and (b) Peltier device. A TE device generally consisted of p- and n-type TE materials connected in series through conducting plates.

$$n_{max} = \frac{T_H - T_C}{T_H} \bullet \frac{\sqrt{1+\overline{ZT}} - 1}{\sqrt{1+\overline{ZT}} + \frac{T_c}{T_h}} \tag{7}$$

$$P_{max} = \frac{(T_H - T_C)^2}{4L} \bullet \overline{S^2\sigma} \tag{8}$$

where L is the length of the TE leg and T_C and T_H are the cold-side and hot-side temperatures, respectively.

The cooling efficiency of a TE cooling device is characterized by the coefficient of performance (COP):

$$COP = \frac{T_C}{T_{H-T_C}} \frac{\sqrt{1+\overline{ZT}} - \frac{T_H}{T_C}}{\sqrt{1+\overline{ZT}} + 1} \tag{9}$$

The term $\overline{ZT}$ [otherwise reported as ZT_M with T_M the average temperature $T_M = (T_H + TC)/2$] is the average value of ZT, the TE device figure of merit, between the hot and cold sides and is defined by Eq. (10):

$$ZT = \frac{S^2}{RK} \bullet T \tag{10}$$

where S, R, K and T are Seebeck coefficient, electrical resistance, thermal conductance and absolute temperature. Another expression of COP is described by Eq. (11):

$$COP = \frac{S\,T_C I - \frac{I^2R}{2} - K\Delta T}{S\,I\,\Delta T + \frac{I^2R}{2}}, \Delta T_{max} = \frac{Z\,T_H^2}{2} \tag{11}$$

where I is the current, R is the resistance and K is the thermal conductance. If 20°C of cooling is required, the COP would typically be in the region of 2. For comparison, a conventional refrigerator under the same circumstances has a COP of around 14. The maximum temperature difference possible for a TE cooler, ΔT_{max}, is often around 50 K.

If the module is used as a heat pump, the COP_{max} of TE heat pump is given by Eq. (12):

$$COP = S\,I\,\Delta T + S\,T_C\,I + \frac{I^2 R}{2} - K\Delta T \qquad (12)$$

As an example, for a temperature difference of 20°C, the *COP* for a TE heat pump would typically be 3, which is comparable to the *COP* of conventional heat pumps.

Besides, the Peltier effect can be used to create a refrigerator that is compact and has no circulating fluid or moving parts. Such refrigerators are useful in applications where their advantages outweigh the disadvantage of their very low efficiency. The Peltier effect is also used by many thermal cyclers, laboratory devices used to amplify DNA by the polymerase chain reaction (PCR). Finally, thermocouples and thermopiles are devices that use the Seebeck effect to measure the temperature difference between two objects. Thermocouples are often used to measure high temperatures, holding the temperature of one junction constant or measuring it independently (cold junction compensation). Thermopiles use many thermocouples electrically connected in series, for sensitive measurements of very small temperature difference.

5. Application of thermoelectric generators (TEGs) for power generation

A TEG is typically used for energy transduction through the Seebeck effect. TEG devices display a variety of advantages compared to other common energy technologies. TEGs function like heat engines but are less bulky and have no moving parts, no noise and long operating lifetime; however, they are typically more expensive and less efficient.

TEGs have wide applications in military, aerospace, cogeneration, medical thermostat, microsensors, etc. They have been used in power plants for converting waste heat into additional electrical power (a form of energy recycling) and in automobiles as automotive thermoelectric generators (ATGs) for increasing fuel efficiency. Space probes often use radioisotope thermoelectric generators (R-TEGs) with the same mechanism but using radioisotopes to generate the required heat difference. Recent uses include body heat—powered lighting and a smartwatch powered by body heat. As examples of TEG applications in the transport sector, e.g. in the aircraft environment, temperature differences can be found in various locations (i.e. between the interior and exterior during flights, near turbines), while in automotive in structural components (e.g. bonnet, between the interior and exterior of the cabin, chassis, exhaust system, etc.). Recently, flexible thermoelectric generators (f-TEGs) have been developed for human body application to power wearable electronic devices with the highest power of 2.28 $\mu W/cm^2$ [111].

6. Recent trends and challenges

6.1 Market of TEGs and recent technologies in TE materials and devices

TEGs for energy harvesting to enable (i) self-powered wireless sensors and (ii) wearable devices have received an exponential growth the last decade. Namely, the annual market for sub-watt TEGs shown in **Figure 7** is expected to grow at a compound annual growth rate of more than 110% over the period 2014 to 2020 [*according to market research firm Infinergia LLC; Grenoble, France*]. In other words the market for these low-power TEGs will be on average more than double each

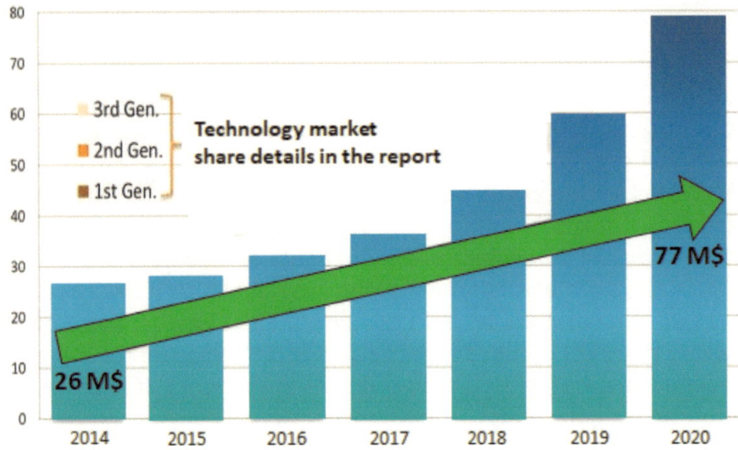

Figure 7.
Annual market size for sub-watt thermoelectric generators (TEGs) with growth at a compound annual growth rate of more than 110% over the period of 2014 to 2020 [according to market research firm Infinergia LLC; Grenoble, France].

year from about 100,000 units shipped in 2014 to about 9 million shipped in 2020. A market size of TEGs considering low-power units is estimated to move from $26 million in 2014 to $77 million in 2020 (thus reaching a CAGR of about 20%).

TEGs are moving into second- and third-generation technologies which are opening up new opportunities, according to Infinergia. Products using sub-watt thermal energy harvesting are being commercialized in several applications across two main segments: (a) infrastructure and buildings and (b) industrial and professional.

The most recent established technology on TE materials and TEGs is related to organic thermoelectrics and organic TEGs (OTEGs). OTEGs can be deposited by vacuum technologies or facile and scalable solution deposition techniques on rigid or flexible substrates (e.g. flexible glass, PET, PEN, etc.). Organic semiconductor materials have advantages of low-cost, lightweight, mechanical flexibility and low-temperature solution processability over large areas, enabling the development of personal, portable and flexible thermal modules. Polymer-based (conducting and nonconducting), nanocarbon-based (CNTs, graphene, fullerenes, etc.) and nanocomposite systems have been reported as TE materials, and in many cases, TEG prototypes have been fabricated to demonstrate their potential for power generation. Critical challenges of organic TE materials include lifetime stability in operation and the relatively low temperatures that can withstand as TE materials. However, the continuous increase of OTEG power output together with a major recent trend of combining TE and photovoltaic devices to maximize the electric energy output [112] comprises a viable avenue for the future market of organic TE materials and OTEGs.

Figure 8 demonstrates a fully printed SWCNT f-TEG (or OTEG) fabricated onto a flexible Kapton polymeric film substrate. Kapton polyimide-based polymer is a high-temperature engineering thermoplastic that exhibits a thermal stability of >500°C [113]. The f-TEG upon being exposed to a temperature gradient of ~110 K creates a thermoelectric voltage output of 41.1 mV. The TEG device demonstrated has been recently fabricated by the sole author of the current book chapter (Dr. Lazaros Tzounis), while more experimental details will be included in scientific publication which is under preparation.

f-TEGs have been reported also based on conventional inorganic low bandgap semiconductor nanomaterials (Bi_2Te_3, PbTe, Ag_2Te_3, etc.) that in the form of

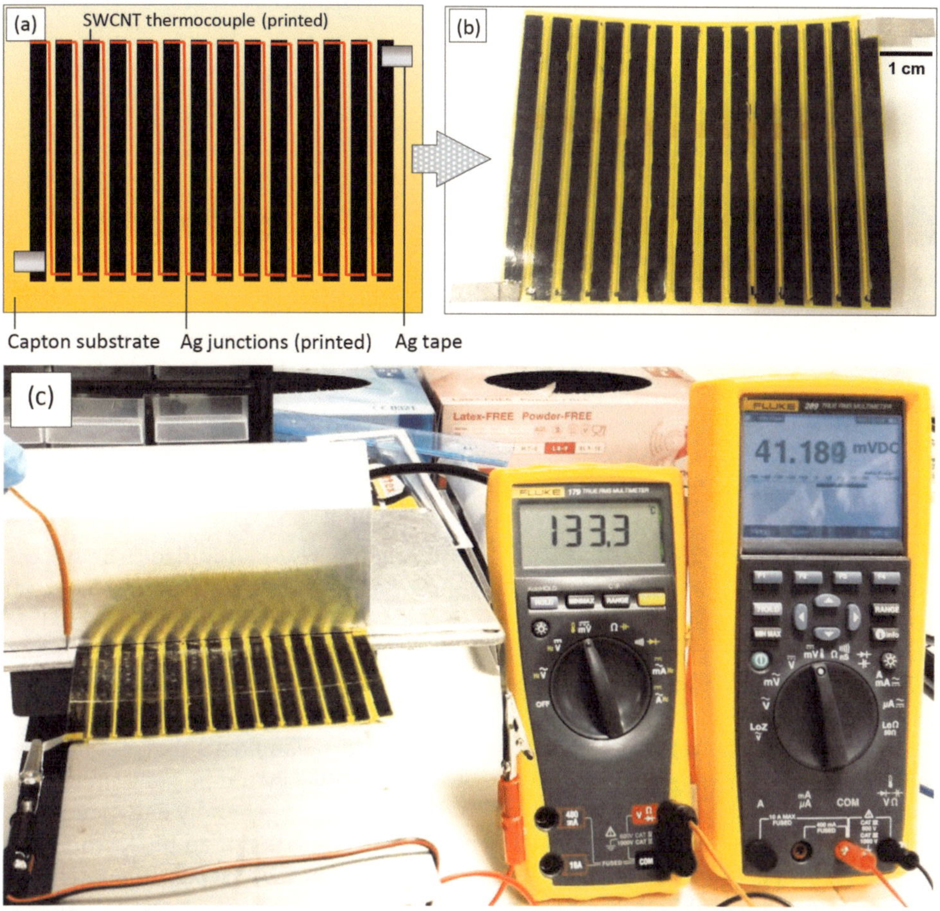

Figure 8.
(a) A carton of a fully printed flexible OTEG with 14 p-type serially interconnected SWCNT thermocouples with Ag junctions, (b) the demonstration of the real device by a digital photo and (c) the f-TEG in operation yielding 41.1 mV upon being exposed to a $\Delta T = \sim 110$ *K.*

colloidal nanocrystal inks (or pastes) can be printed on various flexible substrates enabling highly efficient TEGs. Printing techniques that have been utilized range from slot-die printing [114], inkjet printing, screen printing [115] to aerosol jet printing [116]. Namely, screen printing technique has been deployed by Varghese et al. [117] to fabricate an f-TEG with high figure of merit onto flexible polyimide substrates. First, bismuth telluride-based nanocrystals have been synthesized using a microwave-stimulated wet chemical method and formulated further as inks. N-type-printed films demonstrated a peak *ZT* of 0.43 along with superior flexibility, which is among the highest reported *ZT* values in flexible thermoelectric materials. A flexible TEG-fabricated device using the printed films exhibited a high power density of 4.1mW/cm^2 at 60°C temperature difference. The additive printing can enable a highly scalable and low-cost roll-to-roll (R2R) manufacturing process to transform high-efficiency colloidal nanocrystals into high-performance and flexible TEG devices for various applications.

Figure 9 demonstrates another potential application of printed f-TEGs by their application underneath flexible photovoltaics (e.g. organic and/or perovskite photovoltaics) in order to increase the total overall efficiency of the hybrid PV-TEG resulting devices. In that case, PVs upon operation and due to light absorbance increase dramatically their temperature at the back contact with the temperature

Figure 9.
Demonstration of a hybrid PV-TEG device with an f-TEG coupled with a perovskite or organic photovoltaic towards the increase of the total efficiency.

values rising possibly up to 100–120°C. Therefore, the TEG can utilize this temperature increase as a heat sink and can drive through its other surface-side (cold side) carriers generating Seebeck voltage.

6.2 Challenges in potential TEG application

In order to realize an f-TEG (and any kind of TEG technology) that can power practical devices or for energy harvesting, at least 20 mV should be generated. This value arises from the fact that the voltage output should be generally enhanced via a voltage step-up converter for practical applications (powering of an LED, storage of the harvested energy in a capacitor, etc.). Currently, commercial step-up converters, which are highly integrated DC/DC converters and have no additional power needed, e.g. the LTC3108 (Linear Tech.), operate at inputs of at least 20 mV to give an output voltage of 2.2 V or even higher. Utilizing the LTC3108, Wei et al. [118] powered a light-emitting diode (LED) via a polymer-based f-TEG containing 300 pieces of parallel connected thermocouples (10 in parallel, 30 in series) generating a power output of ~50 μW with an open circuit voltage higher than 40 mV.

Utilizing TEGs with a voltage output of 20–50 mV, with or without a step-up converter driving the DC voltage produced in a capacitor, could enable the powering of low-consumption electronic devices integrated in buildings, e.g. ultra-low-power microcontrollers, wireless sensor networks, etc. [119]. When the TEG produces voltage outputs greater than 100 mV, more sophisticated devices in applications including smart homes and the Internet of things, to name a few, could be powered [120].

7. Future perspectives

There is no doubt that sheet-to-sheet (S2S) printing additive manufacturing technology on planar rigid or flexible substrates will be the most promising technology for TEG fabrication in the future. Printing technology for TEGs is the most rational choice for the fast preparation of thermoelectric films that can be in principle applied, e.g. with ease to hot tubes, whereby one can extract energy from waste heat. The printing techniques can be easily transferred to a high-yield and high-throughput R2R process for scalable and large-scale production of TEGs on flexible substrates.

It is clear that the carrier substrate, which is required for the S2S or R2R processing, leads to a thermal gradient loss as a significant part of the volume is occupied by the carrier. This should be counteracted by employing thinner carrier substrates. A significant improvement should be also considered regarding the maximum possible area usage of the carrier substrate (known also as filling factor—FF).

Related to organic thermoelectric materials and OTEGs, significant improvements by three to four orders of magnitude are required in terms of *ZT*. The sustainable use of organic thermoelectrics from devices with energy payback times comparable with the lifetime of devices is unlikely unless higher thermal gradients can be employed (larger than the temperature range that organic materials can generally endure) [114].

In an attempt to improve the overall energy conversion efficiency, TE devices could be combined with other devices such as solar cells. Many TE materials can be also used in solar cells, such as Si and metal chalcogenides. The wide application of clean solar energy would significantly reduce fossil fuel consumption, our CO_2 footprint, and environmental deterioration, so the use of multiple energy conversion devices to yield maximal output would be an important direction in this area.

It could be envisaged that TEG technology could contribute to (i) the protection of the environment due to the reduction of conventional fuel usage; (ii) strengthen the interfaces between the energy and transport, transmission, and distribution systems; and (iii) promote synergies with the energy/ICT sectors.

8. Summary and conclusions

In this chapter, recent advances in both inorganic and organic TE materials from experimental and theoretical perspectives have been summarized. Detailed *ZT* values for inorganic and organic TE materials were listed in **Table 1**. In most cases, the enhancement of TE performance is attributed to the introduction of nanostructures into the host matrix, including nanoinclusions, nanocomposites, nanowires, nanograins, point defects, etc. Therefore, the development of TE performance depends on the advances in nanoscience and nanotechnology. Nanostructures can significantly reduce thermal conductivity by enhancing the phonon scattering.

The inorganic TE nanomaterials show high performance; however, the following issues have to be solved: (1) large-scale preparation of size, shape and composition; (2) poor stability of nanostructures, as they can be destroyed during compression at high temperature and/or high pressure into pellets; (3) 1D nanowires and 2D nanosheets are limited for large-scale fabrication of TE devices without destroying their nanostructure; and (4) ordered nanostructures are desired, however, not easily achieved in bulk thermoelectrics. Most inorganic TE materials also have issues of high cost or environmental unfriendliness and cannot be processed in large-scale surfaces for TEG fabrication and large-scale thermal energy harvesting (e.g. geothermal energy harvesting, heat exchangers in industrial pipe systems, aeronautics fuselage, etc.).

A great effort has been given in order to highlight the unique potential of organic materials as thermoelectrics and fabricate accordingly OTEG devices. Relatively, carbon materials such as graphene and carbon nanotubes may be promising candidates for further development of lightweight and low-cost polymer composites for TE applications. The generation of thermoelectricity using organic-based structural, engineering polymeric materials (elastomers, thermoplastics, thermosets) that are routinely exposed to high temperatures can represent a breakthrough in high-performance multifunctional material development.

Acknowledgements

L.T. gratefully acknowledges the Bodossaki Foundation for financial support.

Author details

Lazaros Tzounis
Department of Materials Science and Engineering, University of Ioannina, Ioannina, Greece

*Address all correspondence to: ltzounis@cc.uoi.gr

References

[1] Charisiou ND et al. Syngas production via the biogas dry reforming reaction over nickel supported on modified with CeO_2 and/or La_2O_3 alumina catalysts. Journal of Natural Gas Science and Engineering. 2016;**31**:164-183

[2] Charisiou ND et al. Effect of active metal supported on SiO_2 for selective hydrogen production from the glycerol steam reforming reaction. Bioresources. 2016;**11**:10173-10189

[3] Papageridis KN et al. Comparative study of Ni, Co, Cu supported on γ-alumina catalysts for hydrogen production via the glycerol steam reforming reaction. Fuel Processing Technology. 2016;**152**:156-175

[4] Zhang H et al. A novel phase-change cement composite for thermal energy storage: Fabrication, thermal and mechanical properties. Applied Energy. 2016;**170**:130-139

[5] http://thermoelectrics.matsci.northwestern.edu/thermoelectrics/sustainability.html

[6] Bell LE. Cooling, heating, generating power, and recovering waste heat with thermoelectric systems. Science. 2008; **321**(5895):1457

[7] Volz S. Thermal Nanosystems and Nanomaterials. Berlin Heidelberg: Springer; 2009

[8] Seebeck TJ. Magnetic polarization of metals and minerals. Abhandlungen der Königlichen Akademie der Wissenschaften zu Berlin. 1826;**1823**:265

[9] Peltier JC. Nouvelles expériences sur la caloricité des courants électrique. Annales de Chimie et de Physique. 1834; **56**(14):371-386

[10] Vineis CJ et al. Nanostructured thermoelectrics: Big efficiency gains from small features. Advanced Materials. 2010;**22**(36):3970-3980

[11] Terry MT. Electrical and thermal transport measurement techniques for evaluation of the figure-of-merit of bulk thermoelectric materials. In: Thermoelectrics Handbook. Boca Raton, FL: CRC Press; 2005. pp. 23-1-23-20

[12] Boukai AI et al. Silicon nanowires as efficient thermoelectric materials. Nature. 2008;**451**(7175):168-171

[13] Dresselhaus MS et al. New directions for low-dimensional thermoelectric materials. Advanced Materials. 2007;**19**(8):1043-1053

[14] Pichanusakorn P, Bandaru P. Nanostructured thermoelectrics. Materials Science and Engineering: R: Reports. 2010;**67**(2):19-63

[15] Han C, Li Z, Dou S. Recent progress in thermoelectric materials. Chinese Science Bulletin. 2014;**59**(18): 2073-2091

[16] Snyder GJ, Toberer ES. Complex thermoelectric materials. Nature Materials. 2008;**7**:105

[17] Heremans JP et al. Enhancement of thermoelectric efficiency in PbTe by distortion of the electronic density of states. Science. 2008;**321**(5888):554

[18] Rowe DM. CRC Handbook of Thermoelectrics. Boca Raton, FL: CRC Press; 1995

[19] Hicks LD, Dresselhaus MS. Thermoelectric figure of merit of a one-dimensional conductor. Physical Review B. 1993;**47**(24):16631-16634

[20] Hicks LD, Dresselhaus MS. Effect of quantum-well structures on the thermoelectric figure of merit. Physical Review B. 1993;**47**(19):12727-12731

[21] Tzounis L et al. Controlled growth of Ag nanoparticles decorated onto the surface of SiO_2 spheres: A nanohybrid system with combined SERS and catalytic properties. RSC Advances. 2014;**4**(34):17846-17855

[22] Tzounis L et al. Enhancement of P3HT:PCBM photovoltaic shells efficiency incorporating core-shell Au@Ag plasmonic nanoparticles1. Materials Today: Proceedings. 2016; **3**(3):832-839

[23] Kapnopoulos C et al. Fully gravure printed organic photovoltaic modules: A straightforward process with a high potential for large scale production. Solar Energy Materials and Solar Cells. 2016;**144**:724-731

[24] Kapnopoulos C et al. Gravure printed organic photovoltaic modules onto flexible substrates consisting of a P3HT:PCBM photoactive Blend1. Materials Today: Proceedings. 2016; **3**(3):746-757

[25] Tzounis L et al. Perovskite solar cells from small scale spin coating process towards roll-to-roll printing: Optical and morphological studies. Materials Today: Proceedings. 2017;**4**(4, Part B): 5082-5089

[26] Tzounis L et al. Halloysite nanotubes noncovalently functionalised with SDS anionic surfactant and PS-b-P4VP block copolymer for their effective dispersion in polystyrene as UV-blocking nanocomposite films. Journal of Nanomaterials. 2017;**2017**:11

[27] Gioti M et al. Synthesis, characterization and properties of yellow-light-emitting polyethers containing bis(styryl)anthracene units. AIP Conference Proceedings. 2015; **1646**(1):129-137

[28] Borges BGAL et al. Molecular orientation and ultrafast charge transfer dynamics studies on the P3HT:PCBM blend. The Journal of Physical Chemistry C. 2016;**120**(43):25078-25082

[29] Borges BG et al. Surface, interface and electronic properties of F8:F8BT polymeric thin films used for organic light-emitting diode applications. Polymer International. 2018;**67**:691-699

[30] Wode F et al. Selective localization of multi-wall carbon nanotubes in homopolymer blends and a diblock copolymer. Rheological orientation studies of the final nanocomposites. Polymer. 2012;**53**(20):4438-4447

[31] Tzounis L et al. The interphase microstructure and electrical properties of glass fibers covalently and non-covalently bonded with multiwall carbon nanotubes. Carbon. 2014;**73**: 310-324

[32] Tzounis L et al. High performance natural rubber composites with a hierarchical reinforcement structure of carbon nanotube modified natural fibers. Materials & Design. 2014;**58**:1-11

[33] Liebscher M et al. Influence of the viscosity ratio in PC/SAN blends filled with MWCNTs on the morphological, electrical, and melt rheological properties. Polymer. 2013;**54**(25): 6801-6808

[34] Tzounis L et al. CNT-grafted glass fibers as a smart tool for epoxy cure monitoring, UV-sensing and thermal energy harvesting in model composites. RSC Advances. 2016;**6**(60):55514-55525

[35] Tzounis L et al. Thermal energy harvesting for large-scale applications using MWCNT-grafted glass fibers and polycarbonate-MWCNT nanocomposites. AIP Conference Proceedings. 2015;**1646**(1):138-148

[36] Tzounis L et al. Influence of a cyclic butylene terephthalate oligomer on the processability and thermoelectric properties of polycarbonate/MWCNT

nanocomposites. Polymer. 2014;**55**(21): 5381-5388

[37] Polyzoidis CA et al. Improvement of inverted OPV performance by enhancement of ZnO layer properties as an electron transfer layer. Materials Today: Proceedings. 2016;**3**(3):758-771

[38] Tzounis L et al. All-aromatic SWCNT-polyetherimide nanocomposites for thermal energy harvesting applications. Composites Science and Technology. 2018;**156**: 158-165

[39] Tzounis L et al. Fiber yarns/CNT hierarchical structures as thermoelectric generators. Materials Today: Proceedings. 2017;**4**(7, Part 1): 7070-7075

[40] Tsirka K et al. Production of hierarchical all graphitic structures: A systematic study. Journal of Colloid and Interface Science. 2017;**487**:444-457

[41] Terzopoulou Z et al. Effect of MWCNTs and their modification on crystallization and thermal degradation of poly(butylene naphthalate). Thermochimica Acta. 2017;**656**:59-69

[42] Felisberto M et al. Carbon nanotubes grown on carbon fiber yarns by a low temperature CVD method: A significant enhancement of the interfacial adhesion between carbon fiber/epoxy matrix hierarchical composites. Composites Communications. 2017;**3**:33-37

[43] Carrasco PM et al. Thermoset magnetic materials based on poly(ionic liquid)s block copolymers. Macromolecules. 2013;**46**(5):1860-1867

[44] Tzounis L, Logothetidis S. Fe_3O_4@SiO_2 core shell particles as platforms for the decoration of Ag nanoparticles. Materials Today: Proceedings. 2017;**4**(7, Part 1): 7076-7082

[45] Rowe DM. Recent developments in thermoelectric materials. Applied Energy. 1986;**24**(2):139-162

[46] He J, Liu Y, Funahashi R. Oxide thermoelectrics: The challenges, progress, and outlook. Journal of Materials Research. 2011;**26**(15): 1762-1772

[47] Bubnova O, Crispin X. Towards polymer-based organic thermoelectric generators. Energy & Environmental Science. 2012;**5**(11):9345-9362

[48] Sootsman JR, Chung DY, Kanatzidis MG. New and old concepts in thermoelectric materials. Angewandte Chemie International Edition. 2009; **48**(46):8616-8639

[49] Liu W et al. Recent advances in thermoelectric nanocomposites. Nano Energy. 2012;**1**(1):42-56

[50] Li Z et al. Semiconductor nanowires for thermoelectrics. Journal of Materials Chemistry. 2012;**22**(43):22821-22831

[51] Li Z et al. Formation and function of bismuth nanocatalysts for the solution-liquid-solid synthesis of CdSe nanowires. Small. 2008;**4**(10): 1698-1702

[52] Li J-F et al. High-performance nanostructured thermoelectric materials. NPG Asia Materials. 2010;**2**: 152

[53] Du Y et al. Research progress on polymer–inorganic thermoelectric nanocomposite materials. Progress in Polymer Science. 2012;**37**(6):820-841

[54] Gao C, Chen G. Conducting polymer/carbon particle thermoelectric composites: Emerging green energy materials. Composites Science and Technology. 2016;**124**:52-70

[55] Chen G, Xu W, Zhu D. Recent advances in organic polymer

thermoelectric composites. Journal of Materials Chemistry C. 2017;**5**(18): 4350-4360

[56] He R, Schierning G, Nielsch K. Thermoelectric devices: A review of devices, architectures, and contact optimization. Advanced Materials Technologies. 2017;**3**(4): 1700256

[57] Harman TC et al. Nanostructured thermoelectric materials. Journal of Electronic Materials. 2005;**34**(5): L19-L22

[58] Yan H, Sada N, Toshima N. Thermal transporting properties of electrically conductive polyaniline films as organic thermoelectric materials. Journal of Thermal Analysis and Calorimetry. 2002;**69**(3):881-887

[59] Kim GH et al. Engineered doping of organic semiconductors for enhanced thermoelectric efficiency. Nature Materials. 2013;**12**:719

[60] Hong W et al. Thermally driven large N-type voltage responses from hybrids of carbon nanotubes and poly (3,4-ethylenedioxythiophene) with Tetrakis(dimethylamino)ethylene. Advanced Materials. 2015;**27**(43): 6855-6861

[61] Poudel B et al. High-thermoelectric performance of nanostructured bismuth antimony telluride bulk alloys. Science. 2008;**320**(5876):634-638

[62] Du Y et al. Simultaneous increase in conductivity and Seebeck coefficient in a polyaniline/graphene nanosheets thermoelectric nanocomposite. Synthetic Metals. 2012;**161**(23–24): 2688-2692

[63] Meng C, Liu C, Fan S. A promising approach to enhanced thermoelectric properties using carbon nanotube networks. Advanced Materials. 2010; **22**(4):535-539

[64] Wang L et al. Thermoelectric properties of conducting polyaniline/ graphite composites. Materials Letters. 2011;**65**(7):1086-1088

[65] Tomlinson EP et al. Tuning the thermoelectric properties of a conducting polymer through blending with open-Shell molecular dopants. ACS Applied Materials & Interfaces. 2015; 7(33):18195-18200

[66] Yi C et al. Enhanced thermoelectric properties of poly(3,4-ethylenedioxythiophene):poly (styrenesulfonate) by binary secondary dopants. ACS Applied Materials & Interfaces. 2015;**7**(17):8984-8989

[67] Mengistie DA et al. Enhanced thermoelectric performance of PEDOT: PSS flexible bulky papers by treatment with secondary dopants. ACS Applied Materials & Interfaces. 2015;**7**(1): 94-100

[68] Moriarty GP et al. Thermoelectric behavior of organic thin film nanocomposites. Journal of Polymer Science Part B: Polymer Physics. 2013; **51**(2):119-123

[69] Bounioux C et al. Thermoelectric composites of poly(3-hexylthiophene) and carbon nanotubes with a large power factor. Energy & Environmental Science. 2013;**6**(3):918-925

[70] Song H et al. Fabrication of a layered nanostructure PEDOT:PSS/ SWCNTs composite and its thermoelectric performance. RSC Advances. 2013;**3**(44):22065-22071

[71] Anthopoulos TD et al. Air-stable ambipolar organic transistors. Applied Physics Letters. 2007;**90**(12):122105

[72] Boris R et al. Power factor enhancement in solution-processed organic n-type thermoelectrics through molecular design. Advanced Materials. 2014;**26**(21):3473-3477

[73] Schlitz RA et al. Solubility-limited extrinsic n-type doping of a high electron mobility polymer for thermoelectric applications. Advanced Materials. 2014;**26**(18):2825-2830

[74] Shi K et al. Toward high performance n-type thermoelectric materials by rational modification of BDPPV backbones. Journal of the American Chemical Society. 2015; **137**(22):6979-6982

[75] Dazhen H et al. Bismuth interfacial doping of organic small molecules for high performance n-type thermoelectric materials. Angewandte Chemie International Edition. 2016;**55**(36): 10672-10675

[76] Zhang B et al. Promising thermoelectric properties of commercial PEDOT:PSS materials and their Bi_2Te_3 powder composites. ACS Applied Materials & Interfaces. 2010;**2**(11): 3170-3178

[77] Yu C et al. Thermoelectric behavior of segregated-network polymer nanocomposites. Nano Letters. 2009; **9**(3):1283-1283

[78] Choi K, Yu C. Highly doped carbon nanotubes with gold nanoparticles and their influence on electrical conductivity and thermopower of nanocomposites. PLoS One. 2012;7(9): e44977

[79] Hewitt CA et al. Multilayered carbon nanotube/polymer composite based thermoelectric fabrics. Nano Letters. 2012;**12**(3):1307-1310

[80] Poehler TO, Katz HE. Prospects for polymer-based thermoelectrics: State of the art and theoretical analysis. Energy & Environmental Science. 2012;**5**(8): 8110-8115

[81] Yao Q et al. Enhanced thermoelectric performance of single-walled carbon nanotubes/polyaniline hybrid nanocomposites. ACS Nano. 2010;**4**(4):2445-2451

[82] Wang L et al. Preparation and thermoelectric properties of polythiophene/multiwalled carbon nanotube composites. Synthetic Metals. 2013;**181**:79-85

[83] Liang B et al. Fabrication and thermoelectric properties of graphene/ composite materials. Journal of Nanomaterials. 2013;**2013**:5

[84] Bubnova O et al. Optimization of the thermoelectric figure of merit in the conducting polymer poly(3,4-ethylenedioxythiophene). Nature Materials. 2011;**10**(6):429-433

[85] Dubey N, Leclerc M. Conducting polymers: Efficient thermoelectric materials. Journal of Polymer Science Part B: Polymer Physics. 2011;**49**(7): 467-475

[86] Luo J et al. Polypropylene-based melt mixed composites with single walled carbon nanotubes for thermoelectric applications: Switching from p-type to n-type by the addition of polyethylene glycol. Polymer. 2017;**108**: 513-520

[87] Papageorgiou DG et al. β-Nucleated propylene–ethylene random copolymer filled with multi-walled carbon nanotubes: Mechanical, thermal and rheological properties. Polymer. 2014; **55**(16):3758-3769

[88] Tzounis L et al. Shear alignment of a poly(styrene-butadiene-styrene) triblock copolymer/MWCNT nanocomposite. Polymer. 2017;**131** (Supplement C):1-9

[89] Antar Z et al. Thermoelectric behaviour of melt processed carbon nanotube/graphite/poly (lactic acid) conductive biopolymer nanocomposites (CPC). Materials Letters. 2012;**67**(1): 210-214

[90] Liebscher M et al. Influence of the MWCNT surface functionalization on the thermoelectric properties of melt-mixed polycarbonate composites. Composites Science and Technology. 2014;**101**(0):133-138

[91] Liebscher M et al. Thermoelectric energy harvesting with highly conductive CNT-filled polycarbonate composites prepared by melt-mixing. In: 6th International Conference on Carbon Nanoparticle based Composites; Dresden. 2013

[92] Hewitt CA, Carroll DL. The effects of acid treatment on the thermoelectric power of multiwalled carbon nanotubes synthesized by chemical vapor deposition. Chemical Physics Letters. 2013;**580**(Supplement C:67-72

[93] Zhao W et al. Flexible carbon nanotube papers with improved thermoelectric properties. Energy & Environmental Science. 2012;**5**(1): 5364-5369

[94] Tan XJ et al. Thermoelectric properties of ultrasmall single-wall carbon nanotubes. The Journal of Physical Chemistry C. 2011;**115**(44): 21996-22001

[95] Kong WJ et al. Thermoelectric power of a single-walled carbon nanotubes strand. Journal of Physics: Condensed Matter. 2005;**17**(12):1923

[96] Kunadian I et al. Thermoelectric power generation using doped MWCNTs. Carbon. 2009;**47**(3):589-601

[97] Nonoguchi Y et al. Systematic conversion of single walled carbon nanotubes into n-type thermoelectric materials by molecular dopants. 2013;**3**: 3344. DOI: 10.1038/srep0334

[98] Piao M et al. Thermoelectric properties of single walled carbon nanotube networks in polycarbonate matrix. Physica Status Solidi (b). 2013; **250**(8):1468-1473

[99] Piao M et al. Increasing the thermoelectric power generated by composite films using chemically functionalized single-walled carbon nanotubes. Carbon. 2013;**62** (Supplement C):430-437

[100] Dörling B et al. Exploring different doping mechanisms in thermoelectric polymer/carbon nanotube composites. Synthetic Metals. 2017;**225**(Supplement C): 70-75

[101] Xuan Y et al. Thermoelectric properties of conducting polymers: The case of poly(3-hexylthiophene). Physical Review B. 2010;**82**(11):115454

[102] Han S et al. Morphology and thermoelectric properties of graphene nanosheets enwrapped with polypyrrole. RSC Advances. 2014;**4**(55): 29281-29285

[103] Lu Y, Song Y, Wang F. Thermoelectric properties of graphene nanosheets-modified polyaniline hybrid nanocomposites by an in situ chemical polymerization. Materials Chemistry and Physics. 2013;**138**(1):238-244

[104] Toshima N, Imai M, Ichikawa S. Organic-inorganic nanohybrids as novel thermoelectric materials: Hybrids of polyaniline and bismuth(III) telluride nanoparticles. Journal of Electronic Materials. 2011;**40**(5):898-902

[105] Xu K, Chen G, Qiu D. Convenient construction of poly(3,4-ethylenedioxythiophene)-graphene pie-like structure with enhanced thermoelectric performance. Journal of Materials Chemistry A. 2013;**1**(40): 12395-12399

[106] Kim GH, Hwang DH, Woo SI. Thermoelectric properties of nanocomposite thin films prepared with poly(3,4-ethylenedioxythiophene) poly

(styrenesulfonate) and graphene. Physical Chemistry Chemical Physics. 2012;**14**(10):3530-3536

[107] Zhang K, Zhang Y, Wang S. Enhancing thermoelectric properties of organic composites through hierarchical nanostructures. Scientific Reports. 2013; **3**:3448

[108] Moriarty GP, De S, King PJ, Khan U, Via M, King JA, et al. Thermoelectric behavior of organic thin film nanocomposites. Journal of Polymer Science Part B: Polymer Physics. 2013; **51**(2):119-123

[109] Khan ZU et al. Acido-basic control of the thermoelectric properties of poly (3,4-ethylenedioxythiophene)tosylate (PEDOT-Tos) thin films. Journal of Materials Chemistry C. 2015;**3**(40): 10616-10623

[110] See KC et al. Water-processable polymer–nanocrystal hybrids for thermoelectrics. Nano Letters. 2010; **10**(11):4664-4667

[111] Kim CS et al. Structural design of a flexible thermoelectric power generator for wearable applications. Applied Energy. 2018;**214**:131-138

[112] Lee JJ et al. All organic-based solar cell and thermoelectric generator hybrid device system using highly conductive PEDOT:PSS film as organic thermoelectric generator. Solar Energy. 2016;**134**:479-483

[113] McKeen L. 6—Polyimides. In: The Effect of Sterilization on Plastics and Elastomers. 3rd ed. Boston: William Andrew Publishing; 2012. pp. 169-182

[114] Søndergaard Roar R et al. Practical evaluation of organic polymer thermoelectrics by large-area R2R processing on flexible substrates. Energy Science & Engineering. 2013; **1**(2):81-88

[115] Han C et al. High-performance PbTe thermoelectric films by scalable and low-cost printing. ACS Energy Letters. 2018;**3**(4):818-822

[116] Ou C et al. Fully printed organic–inorganic nanocomposites for flexible thermoelectric applications. ACS Applied Materials & Interfaces. 2018; **10**(23):19580-19587

[117] Varghese T et al. High-performance and flexible thermoelectric films by screen printing solution-processed nanoplate crystals. Scientific Reports. 2016;**6**:33135

[118] Wei Q et al. Polymer thermoelectric modules screen-printed on paper. RSC Advances. 2014;**4**(54): 28802-28806

[119] Shi Y et al. A novel self-powered wireless temperature sensor based on thermoelectric generators. Energy Conversion and Management. 2014;**80**: 110-116

[120] Xin L, Shuang-Hua Y. Thermal energy harvesting for WSNs. In: 2010 IEEE International Conference on Systems, Man and Cybernetics. 2010

Chapter 3

Quantum Theory of the Seebeck Coefficient in YBCO

Shigeji Fujita and Akira Suzuki

Abstract

The measured in-plane thermoelectric power (Seebeck coefficient) S_{ab} in YBCO below the superconducting temperature T_c ($\sim$94 K) S_{ab} is negative and T-independent. This is shown to arise from the fact that the "electrons" (minority carriers) having heavier mass contribute more to the thermoelectric power. The measured out-of-plane thermoelectric power S_c rises linearly with the temperature T. This arises from moving bosonic pairons (Cooper pairs), the Bose-Einstein condensation (BEC) of which generates a supercurrent below T_c. The center of mass of pairons moves as bosons. The resistivity ρ_{ab} above T_c has T-linear and T-quadratic components, the latter arising from the Cooper pairs being scattered by phonons.

Keywords: Seebeck coefficient, in-plane thermoelectric power, out-of-plane thermoelectric power, moving bosonic pairons (Cooper pairs), Bose-Einstein condensation, supercurrent, YBCO

1. Introduction

In 1986, Bednorz and Müller [1] reported their discovery of the first of the high-T_c cuprate superconductors (La-Ba-Cu-O, $T_c > 30$ K). Since then many investigations [2, 3] have been carried out on high-T_c superconductors (HTSC) including Y-Ba-Cu-O (YBCO) with $T_c \sim 94$ K [4]. These compounds possess all of the main superconducting properties, including zero resistance, Meissner effect, flux quantization, Josephson effect, gaps in the excitation energy spectra, and sharp phase transition. In addition these HTSC are characterized by (i) two-dimensional (2D) conduction, (ii) short zero-temperature coherence length ξ_0 ($\sim$ 10Å), (iii) high critical temperature T_c ($\sim$ 100 K), and (iv) two energy gaps. The transport behaviors above T_c are significantly different from those of a normal metal.

YBCO has a critical (superconducting) temperature $T_c \sim 94$ K, which is higher than the liquid nitrogen temperature (77 K). This makes it a very useful superconductor. Terasaki et al. [5, 6] measured the resistivity ρ, the Hall coefficient R^H, and the Seebeck coefficient (thermoelectric power) S in YBCO above the critical temperature T_c. A summary of the data is shown in **Figure 1**. In-plane Hall coefficient R^H_{ab} is positive and temperature (T)-independent, while in-plane Seebeck coefficient S_{ab} is negative and T-independent (anomaly). Thus, there are different charge carriers for the Ohmic conduction and the thermal diffusion. We know that the carrier's mass is important in the Ohmic currents. Lighter mass particles contribute more to the conductivity. The T independence of R^H_{ab} and S_{ab} suggests that "electrons" and "holes" are responsible for the behaviors. We shall explain this behavior,

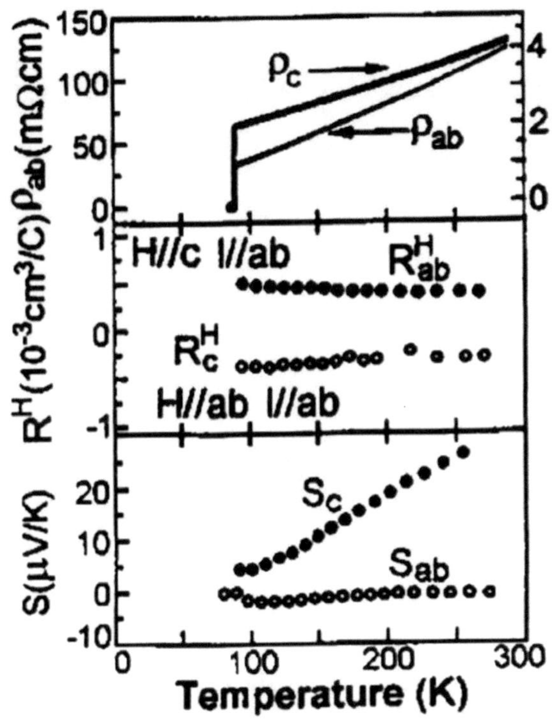

Figure 1.
Normal-state transport of highly oxygenated $YBa_2Cu_3O_{7-\delta}$ after Terasaki et al.'s [5, 6]. Resistivities (top panel); Hall coefficients (middle panel); Seebeck coefficient (bottom panel). The subscripts ab and c denote in-copper plane and out-of-plane directions, respectively.

by assuming "electrons" and "holes" as carriers and using statistical mechanical calculations. Out-of-plane Hall coefficient R_c^H is negative and temperature-independent, while out-of-plane Seebeck coefficient S_c is roughly temperature (T)-linear. We shall show that the pairons, whose Bose condensation generates the supercurrents below T_c, are responsible for this strange T-linear behavior. The in-plane resistivity appears to have T-linear and T-quadratic components. We discuss the resistivity ρ above the critical temperature T_c in Section 6.

In this paper we are mainly interested in the sign and the temperature behavior of the Seebeck coefficient in YBCO. But we discuss the related matter for completeness. There are no Seebeck currents in the superconducting state below the critical temperature ($S = 0$).

2. The crystal structure of YBCO: two-dimensional conduction

HTSC have *layered structures* such that the copper planes comprising Cu and O are periodically separated by a great distance (e.g., $a = 3.88$ Å, $b = 3.82$ Å, $c = 11.68$ Å for YBCO). The lattice structure of YBCO is shown in **Figure 2**. The succession of layers along the c-axis can be represented by CuO–BaO–CuO_2–Y-CuO_2–BaO-CuO–[CuO–BaO–...]. The buckled CuO_2 plane where Cu-plane and O-plane are separated by a short distance as shown is called the *copper planes*. The two copper planes separated by yttrium (Y) are about 3 Å apart, and they are believed to be responsible for superconductivity.

The conductivity measured is a few orders of magnitude smaller along the c-axis than perpendicular to it [7]. This appears to contradict the prediction based on the

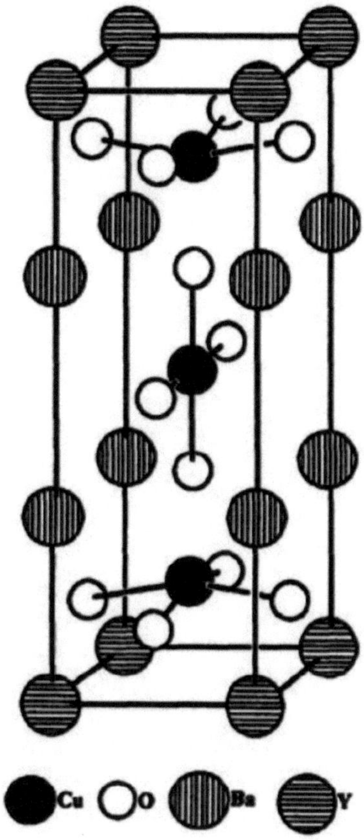

Figure 2.
Arrangement of atoms in a crystal of $YBa_2Cu_3O_7$.

naive application of the Bloch theorem. This puzzle may be solved as follows [8]. Suppose an electron jumps from one conducting layer to its neighbor. This generates a change in the charge states of the layers involved. If each layer is macroscopic in dimension, we must assume that the charge state Q_n of the nth layer can change without limits: $Q_n = \ldots, -2, -1, 0, 1, 2, \ldots$ in units of the electron charge (magnitude) e. Because of unavoidable short circuits between layers due to lattice imperfections, these Q_n may not be large. At any rate if Q_n are distributed *at random* over all layers, then the periodicity of the potential for electron along the c-axis is destroyed. The Bloch theorem based on the electron potential periodicity does not apply even though the lattice is periodic along the c-axis. As a result there are *no* k-vectors along the c-axis. This means that the effective mass in the c-axis direction is infinity, so that *the Fermi surface for a layered conductor is a right cylinder with its axis along the c-axis*. Hence a 2D conduction is established.

Since electric currents flow in the copper planes, there are continuous k-vectors and Fermi energy ε_F. Many experiments [1–3, 9] indicate that a singlet pairs with antiparallel spins called *Cooper pairs* (pairons) form a supercondensate below T_c.

Let us first examine the cause of electron pairing. We first consider attraction via the longitudinal acoustic phonon exchange. Acoustic phonons of lowest energies have long wavelengths λ and a linear energy-momentum (ε-$\hbar k$) relation:

$$\varepsilon = c_s \hbar k, \tag{1}$$

may be assumed, where c_s is the sound speed. The attraction generated by the exchange of longitudinal acoustic phonons is long-ranged. This mechanism is good for a type I superconductor whose pairon size is of the order of 10^4 Å. This attraction is in action also for a HTSC, but it alone is unlikely to account for the much smaller pairon size.

Second we consider the optical phonon exchange. Roughly speaking each copper plane has Cu and O, and 2D lattice vibrations of optical modes are expected to be important. Optical phonons of lowest energies have short wavelengths of the order of the lattice constants, and they have a quadratic dispersion relation:

$$\varepsilon = \varepsilon_0 + A_1\left(k_1 - \frac{\pi}{a_1}\right)^2 + A_2\left(k_2 - \frac{\pi}{a_2}\right)^2, \tag{2}$$

where ε_0, A_1, and A_2 are constants. The attraction generated by the exchange of a massive boson is short-ranged just as the short-ranged nuclear force between two nucleons generated by the exchange of massive pions, first shown by Yukawa [10]. Lattice constants for YBCO are given by $(a_1, a_2) = (3.88, 3.82)$ Å, and the limit wavelengths (λ_{min}) at the Brillouin boundary are twice these values. The observed coherence length ξ_0 is of the same order as λ_{min}:

$$\xi_0 \sim \lambda_{\text{min}} \simeq 8\text{Å}. \tag{3}$$

Thus an electron-optical phonon interaction is a viable candidate for the cause of the electron pairing. To see this in more detail, let us consider the copper plane. With the neglect of a small difference in lattice constants along the a- and b-axes, Cu atoms form a square lattice of a lattice constant $a_0 = 3.85$ Å, as shown in **Figure 3**. Twice as many oxygen (O) atoms as copper (Cu) atoms occupy midpoints of the nearest neighbors (Cu, Cu) in the x_1-x_2 plane.

First, let us look at the motion of an electron wave packet that extends over more than one Cu-site. This wave packet may move easily in $\langle 110 \rangle$ rather than the first neighbor directions $[100]$ and $[010]$. The Bloch wave packets are superposable; therefore, the electron can move in any direction characterized by the two-dimensional k-vectors with bases taken along $[110]$ and $[1\bar{1}0]$. If the number density of electrons is small, the Fermi surfaces should then be a small circle as shown in the central part in **Figure 4**.

Second, we consider a hole wave packet that extends over more than one O-site. It may move easily in $\langle 100 \rangle$ because the Cu-sublattice of a uniform charge

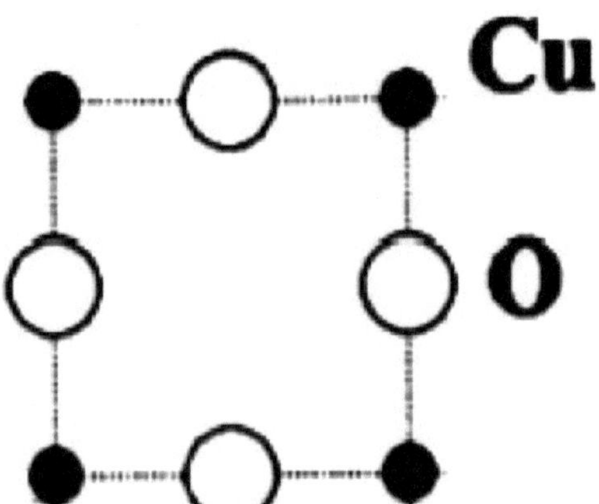

Figure 3.
The idealized copper plane contains twice as many O's as Cu's.

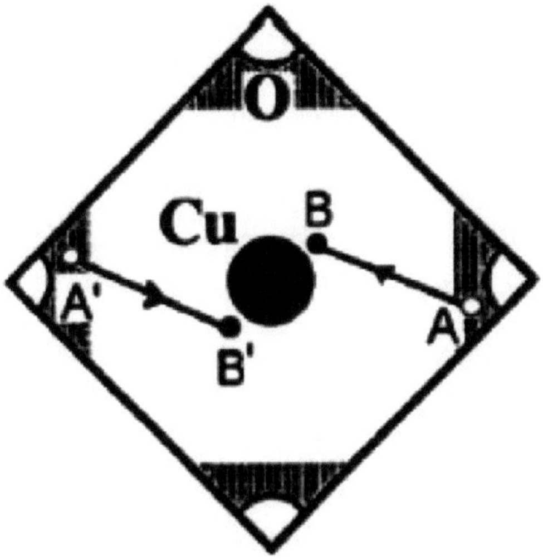

Figure 4.
The two-dimensional Fermi surface of a cuprate model has a small circle (electrons) at the center and a set of four small pockets (holes) at the Brillouin boundary. Exchange of a phonon can create the electron pairon at (B, B') *and the hole pairon at* (A, A')*. The phonon must have a momentum* $p \equiv \hbar k$*, with k being greater than the distance between the electron circle and the hole pockets.*

distribution favors such a motion. If the number of holes is small, the Fermi surface should consist of the four small pockets shown in **Figure 4**. Under the assumption of such a Fermi surface, pair creation of $\pm$ pairons via an optical phonon may occur as shown in the figure. Here a single-phonon exchange generates an electron transition from A in the O-Fermi sheet to B in the Cu-Fermi sheet and another electron transition from A$'$ to B$'$, creating the $-$pairon at (B, B$'$) and the +pairon at (A, A$'$). From momentum conservation the momentum (magnitude) of a phonon must be equal to $\hbar$ times the k-distance AB, which is approximately equal to the momentum of an optical phonon of the smallest energy. Thus an almost insulator-like layered conductor should have a Fermi surface comprising a small electron circle and small hole pockets, which are quite favorable for forming a supercondensate by exchanging an optical phonon.

3. Quantum statistical theory of superconductivity

Following the Bardeen, Cooper, and Schrieffer (BCS) theory [11], we regard the phonon-exchange attraction as the cause of superconductivity. Cooper [12] solved Cooper's equation and obtained a linear dispersion relation for a moving pairon:

$$\varepsilon = w_0 + \frac{1}{2} v_{\mathrm{F}} p, \tag{4}$$

where w_0 is the ground-state energy of the Cooper pair (pairon) and v_{F} is the Fermi speed. This relation was obtained for a three-dimensional (3D) system. For a 2D system, we obtain

$$\varepsilon = w_0 + \frac{2}{\pi} v_{\mathrm{F}} p. \tag{5}$$

The center of mass (CM) motion of a composite is bosonic (fermionic) according to whether the composite contains an even (odd) number of elementary fermions. The Cooper pairs, each having two electrons, move as bosons. In our quantum statistical theory of superconductivity [13], the superconducting temperature T_c is regarded as the Bose-Einstein condensation (BEC) point of pairons. The center of mass of a pairon moves as a boson [13]. Its proof is given in Appendix for completeness. The critical temperature T_c in 2D is given by

$$k_B T_c = 1.24\hbar v_F n^{1/2}, \qquad (6)$$

where n is the pairon density. The inter-pairon distance

$$r_0 \equiv n^{-1/2} = 1.24\hbar v_F (k_B T_c)^{-1} \qquad (7)$$

is several times greater than the BCS pairon size represented by the BCS coherence length:

$$\xi_0 \equiv 0.181\hbar v_F (k_B T_c)^{-1}. \qquad (8)$$

Hence the BEC occurs without the pairon overlap. Phonon exchange can be repeated and can generate a pairon-binding energy ε_b of the order of $k_B T_b$:

$$\varepsilon_b \equiv k_B T_b, \quad T_b \sim 1000\ \text{K}. \qquad (9)$$

Thus, the pairons are there above the superconducting temperature T_c. The angle-resolved photoemission spectroscopy (ARPES) [14] confirms this picture.

In the quantum statistical theory of superconductivity, we start with the crystal lattice, the Fermi surface and the Hamiltonian and calculate everything, using statistical mechanical methods. The details are given in Ref. [15].

Loram et al. [15] extensively studied the electronic heat capacity of $YBa_2CuO_{6+\delta}$ with varying oxygen concentrations $6 + \delta$. A summary of their data is shown in **Figure 5**. The data are in agreement with what is expected of a Bose-Einstein (B-E)

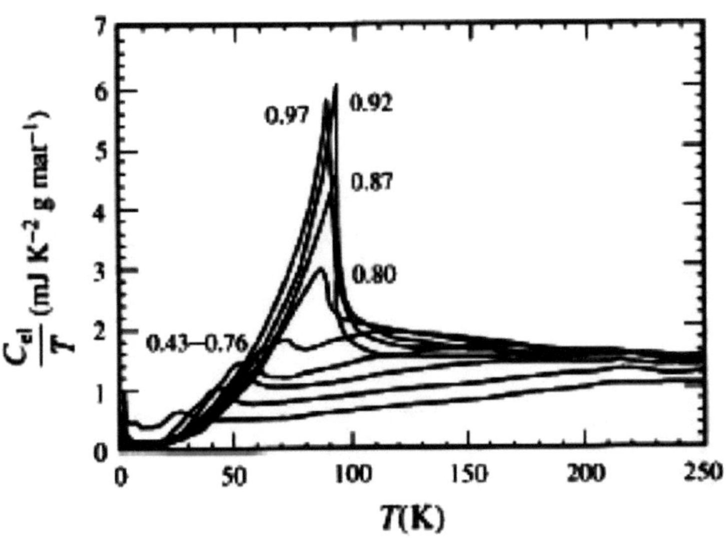

Figure 5.
Electronic heat capacity C_{el} plotted as C_{el}/T vs. temperature T after Loram et al. [15] for $YBa_2Cu_3O_{6+\delta}$ with the δ values shown.

condensation of free massless bosons in 2D, a peak with no jump at T_c with the T^2-law decline on the low-temperature side. The maximum heat capacity at T_c with a shoulder on the high-temperature side can only be explained naturally from the view that the superconducting transition is a macroscopic change of state generated by the participation of a great number of pairons with no dissociation. The standard BCS model regards their T_c as the pair dissociation point and predicts no features above T_c.

The molar heat capacity C for a 2D massless bosons rises like T^2 in the condensed region and reaches $4.38R$ at $T = T_c$; its temperature derivative $\partial C(T, n)/\partial T$ jumps at this point. The *order of phase transition* is defined to be that order of the derivative of the free energy F whose discontinuity appears for the first time. Since $C_V = T(\partial S/\partial T)_V = -T(\partial^2 F/\partial T^2)$, $\partial C_V/\partial T = -T(\partial^3 F/\partial T^3) - (\partial^2 F/\partial T^2)$, the B-E condensation is a *third-order phase transition*. The temperature behavior of the heat capacity C in **Figure 6** is remarkably similar to that of $YBa_2Cu_3O_{6.92}$ (optimal sample) in **Figure** 5. This is an important support for the quantum statistical theory. Other support is discussed in Sections 5 and 6.

Our quantum statistical theory can be applied to 3D superconductors as well. The linear dispersion relation (4) holds. The superconducting temperature T_c in 3D is given by

$$k_B T_c = 1.01 \hbar v_F n^{\frac{1}{3}}, \tag{10}$$

which is identified as the BEC point. The molar heat capacity C for 3D bosons with the linear dispersion relation $\varepsilon = cp$ rises like T^3 and reaches $10.8R$, $R =$ gas constant, at $T_c = 2.02 \hbar c n_0^{1/3}$. It then drops abruptly by $6.57R$ and approaches $3R$ in the high-temperature limit. This temperature behavior of C is shown in **Figure** 7. The phase transition is of second order. This behavior is good agreement with experiments, which supports the BEC picture of superconductivity.

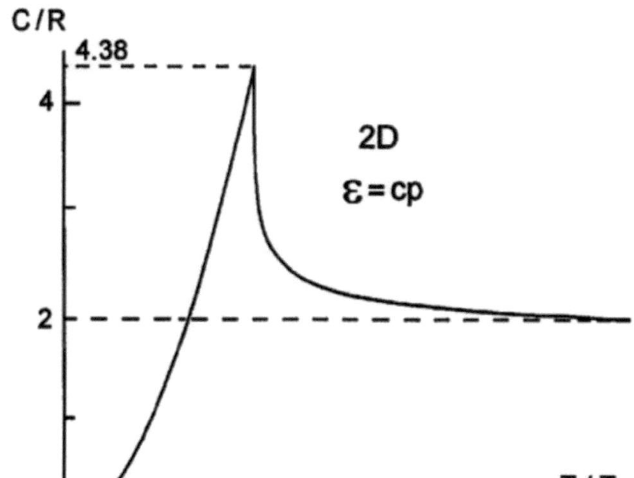

Figure 6.
The molar heat capacity C for 2D massless bosons rise like T^2, reaches $4.38R$ at the critical temperatureT_c, and then decreases to $2R$ in the high-temperature limit.

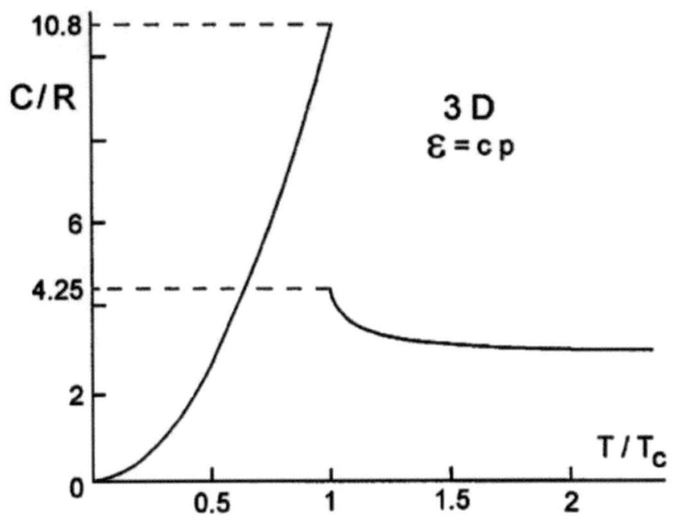

Figure 7.
The molar heat capacity C for 3D massless bosons rises like T^3 and reaches 10.8 R at the critical temperature $T_c = 2.02\hbar c n_0^{1/3}$. It then drops abruptly by 6.57 R and approaches the high-temperature limit 3 R.

4. In-plane Seebeck coefficient above the critical temperature

4.1 Seebeck coefficient for conduction electrons

When a temperature difference is generated and/or an electric field $\boldsymbol{E}$ is applied across a conductor, an electromotive force (emf) is generated. For small potential and temperature gradients, the linear relation between the electric current density $\boldsymbol{j}$ and the gradients

$$\boldsymbol{j} = \sigma(-\nabla V) + A(-\nabla T) = \sigma \boldsymbol{E} - A\nabla T \tag{11}$$

holds, where $\boldsymbol{E} = -\nabla V$ is the electric field and σ is the conductivity. If the ends of the conducting bar are maintained at different temperatures, no electric current flows. Thus from Eq. (11), we obtain

$$\sigma \boldsymbol{E}_S - A\nabla T = 0, \tag{12}$$

where $\boldsymbol{E}_S$ is the field generated by the thermal emf. The Seebeck coefficient S, also called the thermoelectric power or the thermopower, is defined through

$$\boldsymbol{E}_S = S\nabla T, \quad S \equiv A/\sigma. \tag{13}$$

The conductivity σ is always positive, but the Seebeck coefficient S can be positive or negative depending on the materials. We present a kinetic theory to explain Terasaki et al.'s experimental results [5, 6] for the Seebeck coefficient in $YBa_2Cu_3O_{7-\delta}$, reproduced in **Figure 1**.

We assume that the carriers are conduction electrons ("electron," "hole") with charge q ($-e$ for "electron," $+e$ for "hole") and effective mass m^*. At a finite temperature $T > 0$, "electrons" ("holes") are excited near the Fermi surface if the surface curvature is negative (positive) [16]. The "electron" ("hole") is a *quasi-electron* which has an energy higher lower than the Fermi energy ε_F *and* which circulates clockwise (counterclockwise) viewed from the tip of the applied magnetic field vector.

"Electrons" ("holes") are excited on the positive (negative) side of the Fermi surface with the convention that the positive normal vector at the surface points in the energy-increasing direction. The number of thermally excited "electrons" $N_{\rm ex}$, having energies greater than the Fermi energy $\varepsilon_{\rm F}$, is defined and calculated as

$$\begin{aligned} N_{\rm ex} &\equiv \int_{\varepsilon_{\rm F}}^{\infty} \mathrm{d}\varepsilon\, \mathcal{D}(\varepsilon) f(\varepsilon, T, \mu) \approx \mathcal{D}(\varepsilon_{\rm F}) \int_{\varepsilon_{\rm F}}^{\infty} \mathrm{d}\varepsilon \frac{1}{e^{\beta(\varepsilon-\mu)}+1} \\ &\simeq \ln 2\, k_{\rm B} T \mathcal{D}(\varepsilon_{\rm F}), \end{aligned} \tag{14}$$

where $\mathcal{D}(\varepsilon)$ is the density of states. This formula holds for 2D and 3D in high degeneracy. The density of thermally excited "electrons,"

$$n_{\rm ex} = N_{\rm ex}/\mathcal{A}, \quad \mathcal{A} = \text{planer area}, \tag{15}$$

is higher at the high-temperature end, and the particle current runs from the high- to the low-temperature end. This means that the electric current runs toward (away from) the high-temperature end in an "electron" ("hole")-rich material. After using Eqs. (13) and (14), we obtain

$$S = \begin{cases} <0 & \text{for "electrons"} \\ >0 & \text{for "holes"} \end{cases} \tag{16}$$

The Seebeck current arises from the thermal diffusion. We assume Fick's law:

$$\boldsymbol{j} = q\boldsymbol{j}_{\rm particle} = -qD\nabla n_{\rm ex}, \tag{17}$$

where D is the *diffusion constant*, which is computed from the standard formula:

$$D = \frac{1}{\rm d} v\ell = \frac{1}{\rm d} v_{\rm F}^2 \tau, \quad v = v_{\rm F}, \quad \ell = v\tau, \tag{18}$$

where $v_{\rm F}$ is the Fermi velocity and τ the relaxation time of the charged particles. The symbol d denotes the dimension. The density gradient $\nabla n_{\rm ex}$ is generated by the temperature gradient ∇T and is given by

$$\nabla n_{\rm ex} = \frac{\ln 2}{\mathcal{A}\rm d} k_{\rm B} \mathcal{D}(\varepsilon_{\rm F}) \nabla T, \tag{19}$$

where Eq. (14) is used. Using Eqs. (17)–(19) and (11), we obtain the thermal diffusion coefficient A as

$$A = \frac{\ln 2}{2\mathcal{A}} q v_{\rm F}^2 k_{\rm B} \mathcal{D}(\varepsilon_{\rm F}) \tau. \tag{20}$$

We divide A by the conductivity

$$\sigma = nq^2\tau/m^*, \tag{21}$$

and obtain the Seebeck coefficient S [see Eq. (13)]:

$$S \equiv A/\sigma = \ln 2 \frac{k_{\rm B}\varepsilon_{\rm F}}{nq} \frac{\mathcal{D}(\varepsilon_{\rm F})}{\mathcal{A}}, \quad \varepsilon_{\rm F} \equiv \frac{1}{2} m^* v_{\rm F}^2. \tag{22}$$

The relaxation time τ cancels out from numerator and denominator. This result is independent of the temperature T.

4.2 In-plane thermopower for YBCO

We apply our theory to explain the in-plane thermopower data for YBCO. For $YBa_2Cu_3O_{7-\delta}$ (composite), there are "electrons" and "holes". The "holes", having smaller m^* and higher $v_F \equiv (2\varepsilon_F/m^*)^{1/2}$, dominate in the Ohmic conduction and also in the Hall voltage V_H, yielding a positive Hall coefficient R_{ab}^{H} (see **Figure 1**). But the experiments indicate that the in-plane thermopower S_{ab} is negative. This puzzle may be solved as follows.

We assume an effective mass approximation for the in-plane "electrons":

$$\varepsilon = \left(p_x^2 + P_y^2\right)/2m^*. \tag{23}$$

The 2D density of states including the spin degeneracy is

$$\mathcal{D} = m^* A/\left(\pi\hbar^2\right), \tag{24}$$

which is independent of energy. The "electrons" (minority carriers), having heavier mass m_1^*, contribute more to A, and hence the thermopower S_{ab} can be negative as shown below.

When both "electrons" (1) and "holes" (2) exist, their contributions to the thermal diffusion are additive. Using Eqs. (20) and (24), we obtain

$$\begin{aligned} A_{ab} &= -e\ln 2\,\frac{k_B}{2\pi\hbar^2}\left(v_F^{(1)}\right)^2 m_1^*\,\tau_1 + e\ln 2\,\frac{k_B}{2\pi\hbar^2}\left(v_F^{(2)}\right)^2 m_2^*\,\tau_2 \\ &= -e\ln 2\,\frac{k_B\varepsilon_F}{\pi\hbar^2}(\tau_1 - \tau_2). \end{aligned} \tag{25}$$

If phonon scattering is assumed, then the scattering rate is given by

$$\Gamma \equiv \tau^{-1} = n_{ph} v_F s, \tag{26}$$

where s is the scattering diameter and n_{ph} denotes the phonon population given by the Planck distribution function:

$$n_{ph} = \left[\exp\left(\varepsilon_{ph}/k_B T\right) - 1\right]^{-1}, \tag{27}$$

where ε_{ph} is a phonon energy. We then obtain

$$\begin{aligned} \tau_1 - \tau_2 = 1/\Gamma_1 - 1/\Gamma_2 &= \left(n_{ph} v_F^{(1)} s\right)^{-1} - \left(n_{ph} v_F^{(2)} s\right)^{-1} \\ &= \frac{1}{n_{ph} s}\left(\frac{1}{v_F^{(1)}} - \frac{1}{v_F^{(2)}}\right) > 0, \qquad \left(v_F^{(1)} < v_F^{(2)}\right). \end{aligned} \tag{28}$$

The total conductivity is

$$\begin{aligned} \sigma = \sigma_1 + \sigma_2 &= \frac{e^2 n_1}{m_1^*}\tau_1 + \frac{e^2 n_2}{m_2^*}\tau_2 \\ &= \frac{e^2 n_1}{m_1^* v_F^{(1)} n_{ph} s} + \frac{e^2 n_2}{m_2^* v_F^{(2)} n_{ph} s}. \end{aligned} \tag{29}$$

Using Eqs. (25)–(29), we obtain the in-plane thermopower S_{ab} above the critical temperature as

$$S_{\text{ab}} \equiv \frac{A_{\text{ab}}}{\sigma_{\text{ab}}} = -\ln 2 \frac{k_{\text{B}}\varepsilon_{\text{F}}}{\pi e \hbar^2} \left(\frac{1}{v_{\text{F}}^{(1)}} - \frac{1}{v_{\text{F}}^{(2)}} \right) \left(\frac{n_1}{m_1^* v_{\text{F}}^{(1)}} + \frac{n_2}{m_2^* v_{\text{F}}^{(2)}} \right)^{-1}. \tag{30}$$

The factors $n_{\text{ph}}s$ drop out from numerator and denominator. The obtained Seebeck coefficient S_{ab} is negative and T-independent, in agreement with experiments in $\text{YBa}_2\text{Cu}_3\text{O}_{7-\delta}$, reproduced in **Figure 1**.

5. Out-of-plane thermopower

Terasaki et al. [17, 18] and Takenaka et al. [19] measured the out-of-plane resistivity ρ_{c} in $\text{YBa}_2\text{Cu}_3\text{O}_{\text{x}}$. In the range $6.6 < \text{x} < 6.92$, the data for ρ_{c} can be fitted with

$$\rho_{\text{c}} = C_1 \rho_{\text{ab}} + C_2/T, \tag{31}$$

where C_1 and C_2 are constants and ρ_{ab} is the in-plane resistivity. The first term $C_1\rho_{\text{ab}}$ arises from the in-plane conduction due to the (predominant) "holes" and $+$ pairons. The second term C_2/T arises from the $-$ pairons' quantum tunneling between the copper planes [20]. Pairons move with a linear dispersion relation [21]:

$$\varepsilon = \begin{cases} \frac{2}{\pi} v_{\text{F}} p \equiv cp, & p < p_0 \equiv |w_0|/c \\ 0, & \text{otherwise} \end{cases} \tag{32}$$

with $|w_0|$ being the binding energy of a pairon. The Hall coefficient R_c^{H} (current along the c-axis) is observed to be negative, indicating that the carriers have negative charge (see **Figure 1**).

The tunneling current is calculated as follows. A pairon arrives at a certain lattice-imperfection (impurity, lattice defect, etc.) and quantum-jumps to a neighboring layer with the jump rate given by the Dirac-Fermi golden rule

$$w = \frac{2\pi}{\hbar} \left| \langle \boldsymbol{p}_{\text{f}} | U | \boldsymbol{p}_{\text{i}} \rangle \right|^2 \delta(\varepsilon_{\text{f}} - \varepsilon_{\text{i}}) \equiv \frac{2\pi}{\hbar} \mathcal{M}^2 \delta(\varepsilon_{\text{f}} - \varepsilon_{\text{i}}), \tag{33}$$

where $\boldsymbol{p}_{\text{i}}$ ($\boldsymbol{p}_{\text{f}}$) and ε_{i} (ε_{f}) are, respectively, the initial (final) momentum and energy and U is the imperfection-perturbation. We assume a constant absolute squared matrix-elements $\mathcal{M}^2$. The current density $j_c^{(i)}$ along the c-axis due to a group of particles i having charge $q^{(i)}$ and momentum-energy $(\boldsymbol{p}, \varepsilon)$ is calculated from

$$j_{\text{c}}^{(i)} = j_{\text{c, H}}^{(i)} - j_{\text{c, L}}^{(i)} = q^{(i)} a_0 w n^{(i)} \left(v_{\text{H}}^{(i)} - v_{\text{L}}^{(i)} \right), \tag{34}$$

where $n^{(i)}$ is the 2D number density, a_0 the interlayer distance, and $j_{\text{c, H}}^{(i)}$ $\left(j_{\text{c, L}}^{(i)}\right)$ represents the current density from the high (low)-temperature end. Pairons move with the same speed $c = (2/\pi)v_{\text{F}}$, but the velocity component v_x is

$$v_x = \frac{\partial \varepsilon}{\partial p_x} = \frac{c p_x}{p} = \frac{c^2}{\varepsilon} p_x. \tag{35}$$

Lower-energy (smaller p) pairons are more likely to get trapped by the imperfection and going into tunneling. We represent this tendency by $K = B/\varepsilon$, where B is a constant having the dimension of energy/length. Since the thermal average of the v is different, a steady current is generated. The temperature difference $\Delta T (= T_H - T_L)$ causes a change in the B-E distribution F:

$$F(\varepsilon) \equiv \left[e^{(\varepsilon-\mu)\beta} + 1\right]^{-1}, \quad \beta \equiv (k_B T)^{-1}, \tag{36}$$

where μ is the chemical potential. We compute the current density j_c from

$$j_c = 2e \frac{\mathcal{M}^2 B a_0 \Delta T}{\hbar^3 c^2 \, k_B T^2} \int_0^{cp_0} d\varepsilon \frac{dF(\varepsilon)}{d\beta}, \tag{37}$$

assuming a small ΔT. Not all pairons reaching an imperfection are triggered into tunneling. The factor B contains this correction.

At the BEC temperature (T_c), the chemical potential μ vanishes:

$$\mu(T_c) = 0, \tag{38}$$

and

$$\beta\mu \equiv \mu/k_B T \tag{39}$$

is negative and small in magnitude for $T > T_c$. For high temperature and low density, the B-E distribution function F can be approximated by the Boltzmann distribution function:

$$F(\varepsilon) \approx f_0(\varepsilon) = \exp{(\mu - \varepsilon)\beta}, \tag{40}$$

which is normalized such that

$$\frac{1}{(2\pi\hbar)^2} \int d^2p f_0(\varepsilon) = n_0 \text{ (pairon density)}. \tag{41}$$

All integrals in (37) and (41) can be evaluated simply by using $\int_0^\infty dx e^{-x} x^n = n!$. Hence we obtain

$$\int d^2p f_0(\varepsilon) = 2\pi m \int_0^\infty d\varepsilon e^{\beta\mu} e^{-\beta\varepsilon} = 2\pi m e^{\mu\beta} \beta^{-1}. \tag{42}$$

The integral in (37) is then calculated as

$$\int_0^{cp_0} d\varepsilon \frac{dF(\varepsilon)}{d\beta} \approx \int_0^{cp_0} d\varepsilon \frac{d}{d\beta} f_0(\varepsilon) = e^{\mu\beta} \int_0^\infty d\varepsilon \varepsilon e^{-\beta\varepsilon} = e^{\beta\mu} \beta^{-2}. \tag{43}$$

From Eqs. (11) and (37) along with Eq. (43), we obtain

$$A_c \sim 2e\mathcal{M}^2 B k_B a_0 \left(\hbar^3 c^2\right)^{-1}, \tag{44}$$

which is T-independent.

Experiments [5] indicate that the first term $C_1 \rho_{ab}$ in (31) is dominant for x > 6.8:

$$\rho_c \sim C_1 \rho_{ab} \propto T. \tag{45}$$

Hence at x = 7, we have an expression for the out-of-plane Seebeck coefficient S_c above the critical temperature:

$$S_c \equiv \frac{A_c}{\sigma_c} = A_c C_1 \rho_{ab} \propto T > 0. \ (\rho_{ab} \propto T). \tag{46}$$

The lower the temperature of the initial state, the tunneling occurs more frequently. The particle current runs from the low- to the high-temperature end, the opposite direction to that of the conduction in the ab-plane. Hence $S_c > 0$, which is in accord with experiments (see **Figure 1**).

6. Resistivity above the critical temperature

We use simple kinetic theory to describe the transport properties [22]. Kinetic theory was originally developed for a dilute gas. Since a conductor is far from being the gas, we shall discuss the applicability of kinetic theory. The Bloch wave packet in a crystal lattice extends over one unit cell, and the lattice-ion force averaged over a unit cell vanishes. Hence the conduction electron ("electron," "hole") runs straight and changes direction if it hits an impurity or phonon (wave packet). The electron–electron collision conserves the net momentum, and hence, its contribution to the conductivity is zero. Upon the application of a magnetic field, the system develops a Hall electric field so as to balance out the Lorentz magnetic force on the average. Thus, the electron still move straight and is scattered by impurities and phonons, which makes the kinetic theory applicable.

YBCO is a "hole"-type HTSC in which "holes" are the majority carriers above T_c, while $Nd_{1.84}Ce_{0.16}\ CuO_4$ is an "electron"-type HTSC.

6.1 In-plane resistivity

Consider a system of "holes," each having effective mass m_2^* and charge $+e$, scattered by phonons. Assume a weak electric field $\boldsymbol{E}$ applied along the x-axis. Newton's equation of motion for the "hole" with the neglect of the scattering is

$$m_2^* \frac{\mathrm{m}dv_x}{\mathrm{d}t} = eE. \tag{47}$$

Solving it for v_x and assuming that the acceleration persists in the mean-free time τ_2, we obtain

$$v_\mathrm{d} = \frac{eE}{m_2^*}\tau_2 \tag{48}$$

for the drift velocity v_d. The current density (x-component) j is given by

$$j = en_2 v_\mathrm{d} = n_2 \frac{e^2 \tau_2}{m_2^*} E, \tag{49}$$

where n_2 is the "hole" density. Assuming Ohm's law

$$j = \sigma E, \tag{50}$$

we obtain an expression for the electrical conductivity:

$$\sigma_2 = \frac{n_2 e^2}{m_2^*} \frac{1}{\Gamma_2}, \tag{51}$$

where $\Gamma_2 \equiv \tau_2^{-1}$ is the scattering rate. The phonon scattering rate can be computed, using

$$\Gamma_2 = n_{\text{ph}} v_{\text{F}} A_2, \tag{52}$$

where A_2 is the scattering diameter. If acoustic phonons having average energies

$$\langle \hbar\omega_q \rangle \equiv \alpha_0 \hbar\omega_{\text{D}} \ll k_{\text{B}} T, \quad \alpha_0 \sim 0.20 \tag{53}$$

are assumed, then the phonon number density n_{ph} is given by [23].

$$n_{\text{ph}} = n_{\text{a}} [\exp(\alpha_0 \hbar\omega_{\text{D}}/k_{\text{B}}T) - 1]^{-1} \simeq n_{\text{a}} \frac{k_{\text{B}} T}{\alpha_0 \hbar\omega_{\text{D}}}, \tag{54}$$

where

$$n_{\text{a}} \equiv (2\pi)^{-2} \int \mathrm{d}^2 k \tag{55}$$

is the small $\boldsymbol{k}$-space area where the acoustic phonons are located.
Using Eqs. (51), (52), and (54), we obtain

$$\sigma_2 = \frac{C_2 n_2 e^2}{T}, \quad C_2 \equiv \frac{\alpha_0 \hbar\omega_{\text{D}}}{n_{\text{a}} m_2^* k_{\text{B}} v_{\text{F}} A_2}. \tag{56}$$

Similar calculations apply to "electrons." We obtain

$$\sigma_1 = \frac{C_1 n_1 e^2}{T}, \quad C_1 \equiv \frac{\alpha_0 \hbar\omega_{\text{D}}}{n_{\text{a}} m_1^* k_{\text{B}} v_{\text{F}} A_2}. \tag{57}$$

The resistivity ρ is the inverse of the conductivity σ. Hence the resistivity for YBCO is proportional to the temperature T:

$$\rho \equiv \frac{1}{\sigma} \propto T. \tag{58}$$

Let us now consider a system of + pairons, each having charge $+2e$ and moving with the linear dispersion relation:

$$\varepsilon = cp. \tag{59}$$

Since

$$v_x = (\mathrm{d}\varepsilon/\mathrm{d}p)(\partial p/\partial p_x) = c\,(p_x/p), \tag{60}$$

Newton's equation of motion is

$$\frac{p}{c}\frac{\mathrm{d}v_x}{\mathrm{d}t} = \frac{\varepsilon}{c^2}\frac{\mathrm{d}v_x}{\mathrm{d}t} = 2eE, \tag{61}$$

yielding $v_x = 2e(c^2/\varepsilon)Et +$initial velocity. After averaging over the angles, we obtain

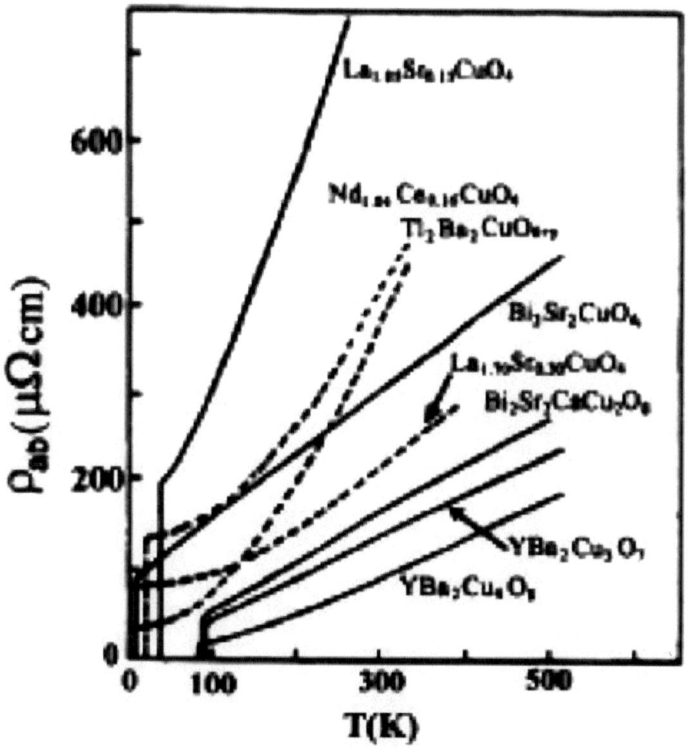

Figure 8.
Resistivity in the ab plane, ρ_{ab} vs. temperature T. Solid lines represent data for HTSC at optimum doping and dashed lines data for highly overdosed samples, after Iye [24].

$$v_{\mathrm{d}}^{(3)} = 2ec^2\tau_3 E\langle\varepsilon^{-1}\rangle, \tag{62}$$

where τ_3 is the pairon mean free time and the angular brackets denote a thermal average. Using this and Ohm's law, we obtain

$$\sigma_3 = (2e)^2 c\langle\varepsilon^{-1}\rangle n_3 \Gamma_3^{-1}, \quad \Gamma_3 \equiv \tau_3^{-1}, \tag{63}$$

where n_3 is the pairon density and Γ_3 is the pairon scattering rate. If we assume a Boltzmann distribution for bosonic pairons above T_c, then we obtain

$$\begin{aligned}\langle\varepsilon^{-1}\rangle &\equiv \left(\frac{2\pi}{(2\pi\hbar)^2}\int_0^\infty dp p\,\frac{1}{\varepsilon}\,e^{\beta cp}\right)\left(\frac{2\pi}{(2\pi\hbar)^2}\int_0^\infty dp p\,e^{\beta cp}\right)^{-1} \\ &= (k_B T)^{-1} \quad \text{for} \quad T > T_c.\end{aligned} \tag{64}$$

The rate Γ_3 is calculated with the assumption of a phonon scattering. We then obtain

$$\sigma_3 = \frac{4n_3 e^2 c^2}{k_B T \Gamma_3} = \frac{2n_2 e^2 C_3}{T^2}, \qquad C_3 \equiv \frac{8}{\pi^2}\frac{\alpha_0 \hbar\omega_D v_F}{n_a k_B^2 A_3}. \tag{65}$$

The total conductivity σ for YBCO is $\sigma_2 + \sigma_3$. Thus taking the inverse of σ, we obtain, by using the results (56) and (65):

$$\rho_{ab} \equiv \frac{1}{\sigma} = \left(\frac{C_2 n_2 e^2}{T} + \frac{C_3 n_3 2e^2}{T^2}\right)^{-1} = \frac{T^2}{n_2 e^2 (C_2 T + 2C_3)} \tag{66}$$

while the conductivity for $Nd_{1.84}Ce_{0.16}CuO_4$ is given by $\sigma_1 + \sigma_3$, and hence the resistivity is similarly given by

$$\rho_{ab} = \frac{T^2}{n_1 e^2 (C_1 T + 2C_3)}. \tag{67}$$

In $Nd_{1.84}Ce_{0.16}$ while in $YCuO_4$ system, "electrons" and − pairons play an essential role for the conduction. In $YBa_2Cu_3O_{7-\delta}$ the "holes" and + pairons are the major carriers in the in-plane resistivity. The resistivity in the plane (ρ_{ab}) vs. temperature (T) in various samples at optimum doping after Iye [24] is shown in **Figure 8**. The overall data are consistent with our formula.

At higher temperature (>160 K), the resistivity ρ_{ab} is linear (see formula (58)):

$$\rho_{ab} \propto T, \quad T > 160\ \text{K}, \tag{68}$$

in agreement with experiments (**Figure 8**). This part arises mainly from the conduction electrons scattered by phonon. At the low temperatures close to the critical temperature T_c, the in-plane resistivity ρ_{ab} shows a T-quadratic behavior [see formula (66)]:

$$\rho_{ab} \propto T^2 \quad (\text{near and above } T_c). \tag{69}$$

This behavior arises mainly from the pairons scattered by phonons. The agreement with the data represents one of the most important experimental supports for the BEC picture of superconductivity.

Author details

Shigeji Fujita[1] and Akira Suzuki[2]*

1 Department of Physics, University at Buffalo, SUNY, Buffalo, NY, USA

2 Department of Physics, Tokyo University of Science, Tokyo, Japan

*Address all correspondence to: asuzuki@rs.kagu.tus.ac.jp

References

[1] Bednorz JG, Müller KA. Zeitschrift für Physik B Condensed Matter. 1986; **64**:189

[2] Halley JW, editor. Theory of High Temperature Superconductivity. Redwood City, CA: Addison-Wesley; 1988

[3] Lundquist S, editor. Towards the Theoretical Understanding of High-T Superconductivity. Singapore: World Scientific; 1988

[4] Wu MK, Ashburn JR, Torng CJ, Hor PH, Meng RL, Gao L, et al. Physical Review Letters. 1987;**58**:908

[5] Terasaki I, Saito Y, Tajima S, Miyamoto S, Tanaka S. Physica C. 1994; **235-240**:1413

[6] Terasaki I, Sato Y, Tajima S, Miyamoto S, Tanaka S. Physical Review B. 1995;**52**:16246

[7] Farrell DE, Beck RG, Booth MF, Allen CJ, Bukowski ED, Ginsberg DM. Physical Review B. 1990;**42**:6758

[8] Godoy S, Fujita S. Journal of Engineering Science. 1991;**29**:1201

[9] Kitazawa K, Ishiguro T, editors. Advances in Superconductivity. Tokyo: Springer; 1989

[10] Yukawa H. Proceedings of the Physico-Mathematical Society of Japan. 1935;**17**:48

[11] Bardeen J, Cooper LN, Schrieffer JR. Physics Review. 1957;**108**:1175

[12] Cooper LN. Physics Review. 1956; **104**:1189

[13] Fujita S, Ito K, Godoy S. Quantum Theory of Conducting Matter, Superconductivity. New York: Springer; 2009

[14] Lanzara A, Bogdanov PV, Zhou XJ, Kellar SA, Feng DL, Lu ED, et al. Nature. 2001;**412**:510

[15] Loram JW, Mirza KA, Cooper JR, Liang WY. Journal of Superconductivity. 1994;**7**:347

[16] Fujita S, Godoy S, Nguyen D. Foundations of Physics. 1995;**25**:1209

[17] Terasaki I, Sato Y, Terajima S. Physical Review B. 1997;**55**:15300

[18] Terasak I, Sato Y, Tajima S. Journal of the Korean Physical Society. 1994; **31**:23

[19] Takenaka K, Mizuhashi K, Takagi H, Uchida S. Physical Review B. 1994;**50**: 6534

[20] Fujita S, Tamura Y, Suzuki A. Modern Physics Letters B. 2000;**14**:30

[21] Fujita S, Godoy S. Theory of High Temperature Superconductivity. The Netherlands: Kluwer Academic; 2001. pp. 65-69, p. 108

[22] Fujita S, Kim SY-G, Okamura Y. Modern Physics Letters B. 2000;**14**:495

[23] Chien TR, Wang ZZ, Ong NP. Physical Review Letters. 1991;**67**:2088

[24] Iye Y. In: Ginzberg DM, editor. Physical Properties of High Temperature Superconductors III. Singapore: World Scientific; 1989

Chapter 4

Bismuth Telluride Solubility Limit and Dopant Effects on the Electronic Properties of Lead Telluride

Dana Ben-Ayoun and Yaniv Gelbstein

Abstract

The demand for energy efficiency has motivated many researchers to seek for novel methods capable of enhancing the conversion of heat to electricity. Most of the recently published methods for thermoelectric (TE) efficiency enhancement discuss on the reduction of the lattice thermal conductivity, with a minor focus on improved electronic optimization. This is attributed mainly to the fact that the electronic properties are correlated and opposing each other upon increasing the carrier concentration. It has been reported that the system of PbTe-BiTe has potentially high TE performance; this chapter is focused on a detailed investigation of the co-effect of bismuth as an effective electronic dopant and at the same time, as a second phase promoter in the PbTe matrix. $(PbTe)_x(BiTe)_{1-x}$ alloys were thermoelectrically examined and the values were analyzed analytically by the general effective media (GEM) approach.

Keywords: thermoelectric, PbTe, BiTe, solubility, GEM

1. Introduction

In the past decades, global climate changes, caused by combustion of fossil fuels and greenhouse gases emission, became a major environmental concern, accompanied with the dilution of conventional energy resources, raising the need for a renewable energy alternatives. Thermoelectricity dealing with this concern, is based on a direct conversion of waste heat into usable electrical energy; even a partial conversion of this waste heat will get us one step closer toward a cleaner and greener world. This goal has been achieved by thermoelectric converters and successfully initiated by the development of various highly efficient thermoelectric material classes. Such materials require a unique combination of the electrical and lattice properties (Seebeck coefficient (α), electrical resistivity (ρ), electronic thermal conductivity (κ_e), and lattice thermal conductivity (κ_l)), enabling the highest possible thermoelectric figure of merit values $ZT = \alpha^2 T/[\rho(\kappa_e + \kappa_l)]$, where T is the absolute temperature, for achieving significant conversion efficiencies. Since the electronic properties are strongly interdependent and follow opposite trends (α and ρ are decreased, and κ_e is increased) upon increasing the carrier concentration, most of the published studies were mainly focused on applying advanced nano-structuring approaches for κ_l reduction.

In the case of IV–VI-based chalcogenides and their alloys (known as among the most efficient thermoelectric alloys for intermediate working temperatures of up to 600°C) such nano-structuring approaches resulted in a significant increase of ZT up to ~2.5, due to an effective scattering of phonons without adversely affecting the electronic properties [1]. Yet, it is worth mentioning that such approaches already succeeded in reaching ultralow thermal conductivity [2], lightening the fact that other approaches, related to electronic optimization of chalcogenides for further enhancement of ZT, are still required.

Among the IV–VI-based chalcogenides, lead telluride is one of the most commonly used thermoelectric materials, but the maximal ZT achieved is only ~0.8 [3] which is still insufficient for converters in widespread industrial use. While thermoelectric properties are dependent on the carrier concentration, materials with carrier concentration higher than 10^{19} cm^{-3} are required. For undoped lead telluride, the change in the carrier concentration and carrier type is obtained by a method that enables the stoichiometry to be changed through thermal annealing in a Pb-rich (for n-type conductivity) or Te-rich (for p-type conductivity). One of the key methods for optimizing the electronic properties of these materials is by adding doping elements to the melt, while the most universally recognized lead telluride dopant is bismuth; it allows obtaining such carrier concentrations that can provide applicative values of thermoelectric conversion efficiency. The thermoelectric properties of the compounds in the system have been investigated, the processes taking place upon increasing the dopant concentration was also studied, but such researches were mainly focused on compositions within the solid solution up to the theoretical solubility limit [4, 5].

This present work, considering our synthesis process, investigates the bismuth effects in the matrix, beneath and beyond the solubility limit, as an effective electronic dopant yet at the same time, as a second phase promoter in the PbTe matrix. Moreover, to the best of our knowledge, most of the previous researches did not investigated the individual electronic contributions of the involved secondary phases embedded in the matrix on the effective thermoelectric transport properties. The general effective media (GEM) approach [6] enables to estimate the transport properties based on experimentally measured properties of each of the involved phases. Since the approach is taking into consideration geometrical aspects, it can be utilized for maximizing the thermoelectric figure of merit of composite materials by artificial alignment of the embedded secondary phase in the composite.

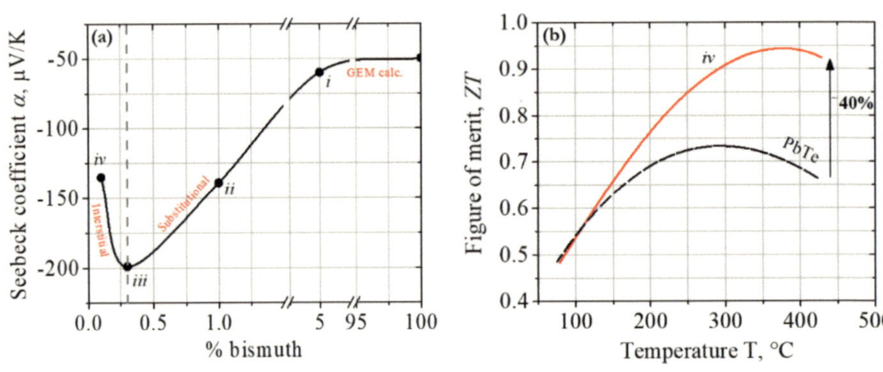

Figure 1.
(a) Bismuth concentration dependence of the room temperature Seebeck coefficient, α, scheme demonstrating the interstitial to substitutional pattern dissolution limit (at ~0.3%) which continues with the GEM calculations for two phase composites consisting an increasing amount of embedded metallic secondary phase. (b) The enhancement of the figure of merit, ZT, in the system along most of the temperature range required for practical operation.

Figure 1a shows the high potential of compositions with low amount of bismuth content, in the vicinity of the solubility limit, to attain high absolute Seebeck coefficient values. **Figure 1b**, demonstrates the potential to enhance the thermoelectric conversion efficiency while considering the co-effect of bismuth as an electronic donor dopant and as a secondary phase promoter in PbTe.

2. Experimental

Three $(PbTe)_x(BiTe)_{1-x}$ alloys, with different x values (**Table 1**), were synthesized from pure elements (5 N), mixed in the right stoichiometric ratio, and sealed in evacuated quartz ampoules under vacuum of 10^{-6} Torr. The ampoules were placed in a rocking furnace (Thermcraft Inc., Winston Salem, NC, USA) at 1000°C for 15 minutes, then water quenched. The cast ingots were milled to a maximal powder particle size of ~250 μm using agate mortar and pestle. The sieved powder was hot pressed (HPW5 Hot Press, FCT System GmbH, Rauenstein, Germany) under a mechanical pressure of 21 MPa at 730°C for 30 minutes under argon atmosphere, resulting in high density values of >98% of the theoretical density.

The thermoelectric transport properties of each alloy were measured up to 450°C as follows. The Seebeck coefficient, α, and electrical resistivity, ρ, were determined using the four-point probe method (Linseis LSR-3/800 Seebeck coefficient/ electrical resistance measuring system). The thermal diffusivity, γ, was determined using the flash diffusivity method (LFA 457, NETZSCH). The total thermal conductivity, κ, was calculated using $\kappa = \gamma \cdot C_P \cdot \delta$, where C_P is the specific heat which was determined using differential scanning calorimetry (DSC 404, NETZSCH), and δ is the density of the sample measured using Archimedes method.

The crystal structure of the alloys was analyzed by X-ray powder diffraction (Rigaku DMAX 2100 powder diffractometer). The microstructure of the alloys was observed using scanning electron microscopy (SEM, JSM-5600, JEOL, Akishima, Japan) equipped with a backscattered electron detector. The chemical composition was measured using an energy-dispersive X-ray spectroscopy (EDS).

Alloy	Bi [%at]	Matrix
i	5%	$(PbTe)_{0.95}(BiTe)_{0.05}$
ii	1%	$(PbTe)_{0.99}(BiTe)_{0.01}$
iii	0.3%	$(PbTe)_{0.997}(BiTe)_{0.003}$
iv	0.1%	$(PbTe)_{0.999}(BiTe)_{0.001}$

Table 1.
The investigated alloys notations.

3. Results and discussion

PbTe has a NaCl cubic crystal structure with space group Fm-3m, melts congruently at 924°C [7]; BiTe has a BiSe trigonal crystal structure with space group P-3m1, consists of 12 close-packed layers along the c axis, melts at 540°C [8].

As can be seen by the XRD diffractogram presented in **Figure 2a**, the reflections are all belong to the lead telluride matrix in the rock salt structure, with no evidence of other phase's peaks. A reason for that might be that the amount of the second phase precipitants detected in the SEM analysis (**Figure 3a** and **b**) is clearly less than the detection limit of the XRD analysis.

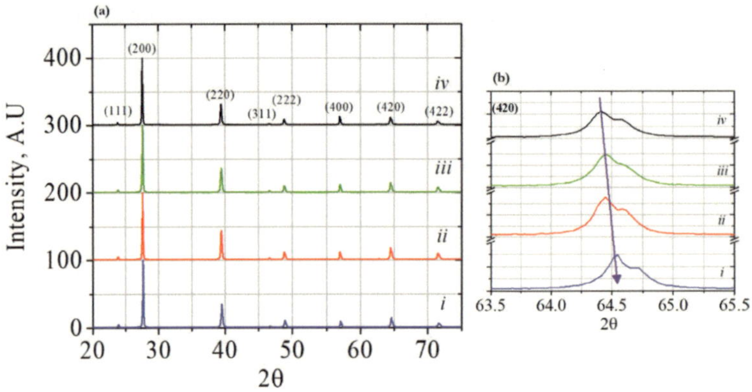

Figure 2.
(a) XRD diffractogram of the investigated i, ii, iii, and iv alloys and (b) a magnification of the (420) reflection at ~64°, indicating a volume decrease upon increasing the bismuth content.

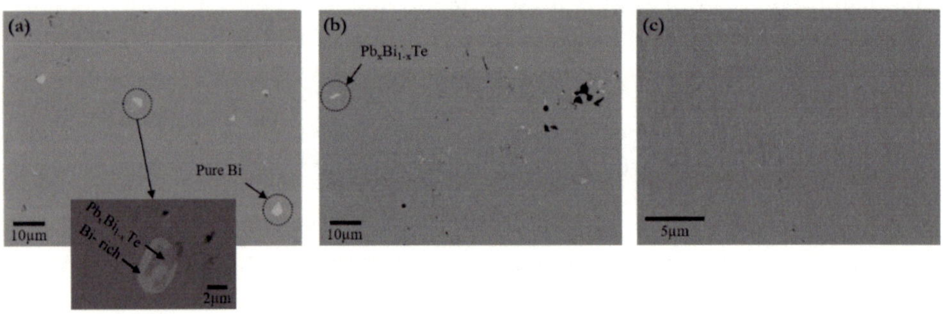

Figure 3.
BSE-SEM micrographs of the investigated (a) i, (b) iii, and (c) iv alloys indicating in the inset the compositional modulations.

As shown in **Figure 2b**, with increasing the bismuth content in the lead telluride matrix, the peaks are constantly shifted toward higher angles, reflecting a constant decrease in the lattice parameter. This may be related to the fact that Bi^{3+} has lower ionic radius (1.03 A [9]) compared to the ionic radius of Pb^{2+} (1.19 A [9]), while bismuth substitutes lead in the NaCl structure. Moreover, it is well known that introducing bismuth to the PbTe matrix creates cation vacancies, which might also contribute to the decrease in lattice parameter while releasing internal stresses/lattice strains. Although broadening of the peaks due to lattice internal stresses caused by bismuth interstitial atoms would be expected, it was not observed in the investigated samples, probably due to the fact that the amount of bismuth at alloy *iv* is too small to affect the lattice.

Back-scattered electrons SEM micrographs of the investigated alloys are presented in **Figure 3a–c**. In **Figure 3a** of the investigated alloy *i*, compositional modulations reveal bright bismuth precipitants along with precipitants composed of several compositions. Since the contrast in back-scattered electrons SEM micrographs is correlated directly to its atomic number, it is obvious that the brighter the phase, the more bismuth it contains; thus, the precipitants contain a PbBiTe ternary phase surrounded by a brighter bismuth-rich phase. In alloy *iii*, PbBiTe ternary phase precipitant can be also detected (**Figure 3b**). The exact composition of the ternary phases, in both samples, could not be defined using EDS analysis. For alloy *iv*, a single-phase matrix, without any precipitant was detected, indicating a full solubility of the elements as can be seen in the representative micrograph in **Figure 3c**.

The logic behind this is that, at alloy *iv*, the dissolution pattern is by interstitial occupation mechanism. It is assumed that bismuth occupation in interstitial sites is less preferred with increasing the bismuth content (*iii*) → (*i*), and the dissolution pattern changes to a substitutional pattern (in line with XRD diffractograms shift, described above); the tendency to participate as a second phase is growing and its effect on the transport properties can be explained by the percolation theory as described by Rogacheva et al. [10]. Upon increasing the bismuth content, percolation passages are formed, causing spatial redistribution of the impurity atoms by self-interaction. By further introduction of bismuth to the matrix, bismuth clusters are formed. Taking the above into consideration, it is reasonable to assume that with increasing the amount of bismuth impurity in the matrix, the probability to form pure bismuth precipitants is growing. Moreover, the high number of ternary compounds existing in bismuth rich PbBiTe system [11], is an indication of a relatively easy restructuring of the lattice. We assume that the precipitant modulations observed might be a result of complexes formation whose composition corresponds to the composition of intermediary phases [10]. Overall, the system is influenced by the percolation effects, and goes through intermediate processes of complex and intermediate structure formations, that can give a reasonable explanation for the observations.

The Seebeck coefficient α, electrical resistivity ρ, thermal conductivity κ and figure of merit ZT, for the investigated alloys are presented in **Figure 4a–d**.

The Bi^{3+} tendency to substitute Pb^{2+} creates one free electron, a fact which obviously affect the electronic properties of the materials. The Seebeck coefficient at room temperature for alloy *i* (−55 μVK^{-1}) indicates the large amount of bismuth introduction to the PbTe matrix; the precipitants acting as a secondary metallic phase, enhancing the carrier concentration. Decreasing the bismuth concentration (alloys *ii* and *iii*) resulted in a gradual improvement of the Seebeck coefficient (−140 and −200 μVK^{-1}, respectively); a dramatic expected change when

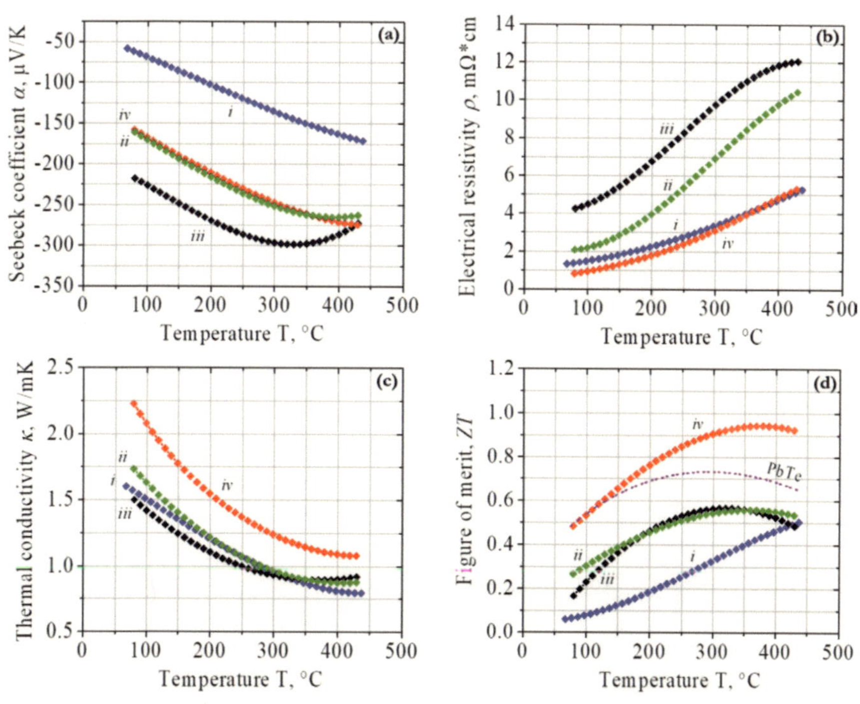

Figure 4.
Temperature dependence of (a) Seebeck coefficient, (b) electrical resistivity, (c) thermal conductivity, and (d) the thermoelectric figure of merit for the investigated i, ii, iii, and iv alloys.

considering the decreased metallic secondary phase amount. By further decreasing the bismuth content (alloy *iv*), the absolute Seebeck coefficient decreased (−135 μVK^{-1}). As can be seen in **Figure 3c**, there is full solubility of bismuth in the PbTe matrix; considering that fact, the reason for the unexpected behavior of Seebeck coefficient might be plant in the dissolution pattern changes as a function of bismuth content. As described above, at low amounts of bismuth, as in alloy *iv*, the dissolution pattern changed to interstitial pattern, where the bismuth electronic activity is relatively low. In that context, a maximum in the absolute Seebeck coefficient values can be reached in the vicinity of the bismuth concentration of alloy *iii*, where the threshold for the substitutional dissolution pattern is reached and it changes to interstitial pattern.

A similar trend observed in **Figure 4b**, from (*i*) → (*iii*), can be seen in the electrical resistivity values. This behavior supports the theory above and is attributed to the increased metallic contribution of the precipitates. Yet, one must keep in mind that bismuth impurities and precipitants cause defect formation in the crystal lattice which consequently scatter the charge carriers and increase the electrical resistivity. While comparing alloy *ii* and *iv*, the increased electrical resistivity in alloy *ii* is attributed to the low mobility due to charge carriers scattering. Moreover, the overlapping between the curves of alloys *i* and *iv* is attributed to the fact that in alloy *i* the charge carrier concentration is relatively high compared to alloy *iv*, while at the same time, in alloy *i* the mobility of the charge carriers is relatively low compared to alloy *iv*. In the last case, these two effects compensate each other and overall, they are affected by the bismuth both as a donor and as a precipitant initiator.

Since the thermal conductivity is affected both by bismuth as an effective electronic donor, but also as a phonon scatter source, both contributions were analyzed. The electronic contribution to the thermal conductivity, as shown in **Figure 5a**, was analyzed using Wiedemann-Franz relation, $\kappa_e = L\,\rho^{-1}\,T$, where L is the Lorentz constant equal to 2.45 10^{-8} WΩK^{-2} calculated in a previously described procedure [12], ρ is the measured electrical resistivity (**Figure 4b**), and T is the absolute temperature. κ_e values follow the expected similar but opposite trends to the ρ values. The lattice contribution to the thermal conductivity, as shown in **Figure 5b**, was calculated by subtracting the calculated κ_e out of the measured κ_{tot} (**Figure 4c**); the gradual change in the κ_l as a function of bismuth content is associated with the cation vacancies described before for *iv*, and continues with increased lattice defects, created by the gradually phase separation observed in **Figure 3a–b**, acting as a phonon scattering source. The overlapping curves can be explained by the fact that the phase separation shown in *iii* (**Figure 3b**) is minor thus might affect κ_l in a very small manner, within

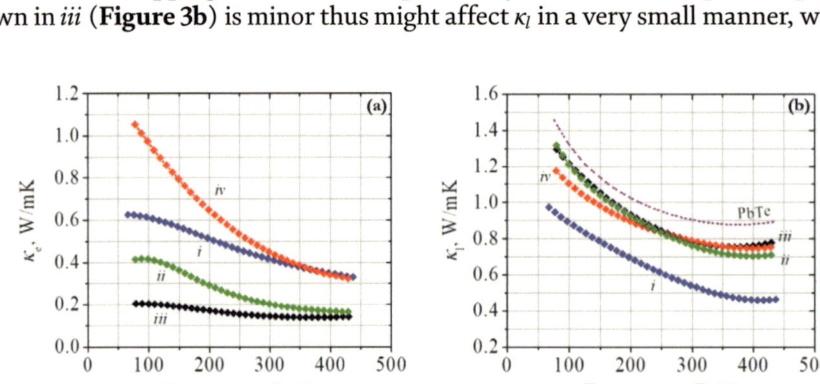

Figure 5.
Temperature dependence of the (a) electronic contribution and (b) lattice contribution to the thermal conductivity of the investigated alloys.

the measurement error of the curves. Overall the results indicate a reduction of ~20 and ~44% upon increased doping compared to pristine PbTe.

It is noted that the dominant thermal conduction mechanism is a bit different between the alloys. While alloys *i* and *iv* are affected by both lattice and charge carrier thermal conductions in the same manner, since their contributions are in the same order of magnitude, alloys *ii* and *iii* charge carrier thermal conduction is very low. This fact opens a window of opportunities for a possible further ZT enhancement by reducing the precipitants (observed in **Figure 3b**) size into the nano-scale for reduction of κ_l.

While combining both the thermal and electrical contributions to the thermoelectric efficiency (**Figure 4d**) it can be concluded that at 450°C, ~40% improvement compared to pristine PbTe was obtained; improvement that is attributed to an optimization between both bismuth contributions as an effective electronic dopant and as a second phase initiator.

In order to further investigate the electronic contribution of the secondary phase due to the geometrical morphology and amount, the general effective media (GEM) approach was applied [6]. The effective thermoelectric transport properties, for composite materials consisting of two separate phases, were calculated based on the effective thermal conductivity, κ_{eff}, and effective electrical resistivity, ρ_{eff}, calculated using Eq. (1) and the effective Seebeck coefficient, α_{eff}, using Eq. (2) [13].

$$x_1 \cdot \frac{(\sigma_1, \kappa_1)^{\frac{1}{t}} - (\sigma_{eff}, \kappa_{eff})^{\frac{1}{t}}}{(\sigma_1, \kappa_1)^{\frac{1}{t}} + A \cdot (\sigma_{eff}, \kappa_{eff})^{\frac{1}{t}}} = (1 - x_1) \cdot \frac{(\sigma_{eff}, \kappa_{eff})^{\frac{1}{t}} - (\sigma_2, \kappa_2)^{\frac{1}{t}}}{(\sigma_2, \kappa_2)^{\frac{1}{t}} + A \cdot (\sigma_{eff}, \kappa_{eff})^{\frac{1}{t}}} \tag{1}$$

$$\frac{\alpha_{eff} - \alpha_2}{\alpha_1 - \alpha_2} = \frac{\frac{\frac{\kappa_{eff}}{\kappa_2}}{\frac{\sigma_{eff}}{\sigma_2}} - 1}{\frac{\frac{\kappa_1}{\kappa_2}}{\frac{\sigma_1}{\sigma_2}} - 1} \tag{2}$$

Where α_1, ρ_1, κ_1 represent the secondary phase embedded in the matrix; α_2, ρ_2, κ_2, represent the matrix; x_1 and $(1 - x_1)$ represent the relative amounts of each of the phases; t represent the homogenous level of distribution of the secondary phase in the matrix, while $t = 1$ stands for an even distribution [14, 15]; A represents the morphological alignment and ranges from "series" (for $A = 0$) to "parallel" (for $A = \infty$) relative to the electrical potential or temperature gradients, while $A = 2$ corresponds to a spherical morphology of the secondary phase [14, 15]. By applying the GEM approach to the investigated case of an embedded bismuth secondary phase inside alloy *iii* matrix, the applied transport properties were the reported values for pure bismuth (−50 μVK^{-1} [16], 0.129 mΩ cm, and 7.97 W mK^{-1} at room temperature [17]) and the experimental data for alloy *iii* at room temperature (**Figure 4a–c**). A was chosen to be equal to different morphological alignment conditions, represented by 0, 2, 8 and infinity.

Using the above data, the effective thermoelectric transport properties, shown in **Figure 6a–d**, were calculated. The measured room temperature experimental values for our *i* and *ii* investigated alloys, are shown in **Figure 6** as black stars; taking into consideration the 10% measurement error, they follow the curve which represent infinitesimal A and uneven distribution of the phase ($t \neq 1$). The conclusion is in line with **Figure 3a–b**, where the precipitants location is mainly in the grain boundary, forcing an elongated shape for the most of them.

As been presented in the experimentally measured values (**Figure 4a–c**), and is in agreement with previously reports, in low amounts of bismuth in the matrix the transport properties are acting abnormally, with regard to the changing pattern of dissolution, up to a point where the dissolution pattern remains steady. Above that

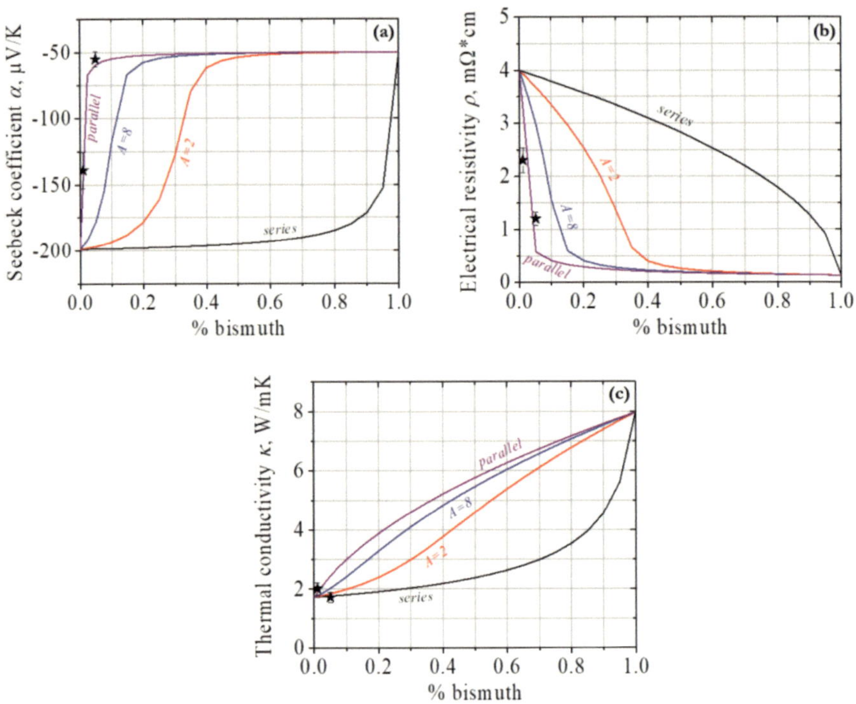

Figure 6.
Calculated GEM curves at room temperature of (a) effective Seebeck coefficient, (b) effective electrical resistivity and (c) effective thermal conductivity as a function of the embedded bismuth percentage, under different morphological alignment and uneven distribution conditions (t = 0.5).

level, the GEM evaluation can predict the transport properties in a very reliable way, since the transport properties of the composite are expected to act as a composite with an increasing amount of a secondary metallic phase.

In our homogeneous distribution level and morphological alignment conditions, it can be seen that up to 0.1%, a slight change in bismuth concentration is accompanied with a drastic change in the electronic properties of the composite. A drastic degradation in the absolute Seebeck coefficient along with a great improvement in the electrical conductivity, sums up in an increased power factor ($\alpha^2\rho^{-1}$); yet while considering the thermal conductivity increment, the overall room temperature thermoelectric efficiency is decreased. On the other hand, the "series" morphological alignment condition at room temperature, exhibits more moderate changes that might allow bigger introduction of bismuth to the matrix, which will reduce the lattice contribution to the thermal conductivity, and at the same time will not affect as drastically the electronic properties. This finding implies on a potential method for enhancing the thermoelectric efficiency by an artificial "series" morphological alignment of the secondary phase which can lead to possible further study.

The reason for the small mismatches observed in our GEM calculations, might be related to the fact that the matrix was represented by alloy *iii*; it is well known that even a slight change in impurity content, affect substantially the transport properties.

4. Conclusions

The co-effect of bismuth as an effective electronic dopant and at the same time, as a second phase promoter in the PbTe matrix was investigated and explained in details with regard to the dissolution pattern transition. These two effects on the thermoelectric properties were demonstrated and resulted in a ~40% ZT enhancement compared to the pristine PbTe. The GEM analysis of composites with an increased amount of precipitates paved a potential route for further enhancement of the thermoelectric efficiency upon an artificial serial alignment of the embedded secondary phase in the composite.

Acknowledgements

The work was supported by the Israel Science Foundation (ISF) Individual Research Grant No. 455/16. The authors would like to thank Mr. Yair George for the synthesis of the alloys and specimens preparation.

Conflict of interest

The authors declare no conflict of interest.

Author details

Dana Ben-Ayoun* and Yaniv Gelbstein
Department of Materials Engineering, Ben-Gurion University of the Negev, Beer-Sheva, Israel

*Address all correspondence to: danabenayoun@gmail.com

References

[1] Hazan E, Madar N, Parag M, Casian V, Ben-Yehuda O, Gelbstein Y. Effective electronic mechanisms for optimizing the thermoelectric properties of GeTe-rich alloys. Advanced Electronic Materials. 2015;**1**(11):1-7

[2] Xu B, Feng T, Agne MT, Zhou L, Ruan X, Snyder GJ, et al. Highly porous thermoelectric nanocomposites with low thermal conductivity and high figure of merit from large-scale solution-synthesized $Bi_2Te_{2.5}Se_{0.5}$ hollow nanostructures. Angewandte Chemie, International Edition. 2017;**56**(13):3546-3551

[3] Komisarchik G, Gelbstein Y, Fuks D. Solubility of Ti in thermoelectric PbTe compound. Intermetallics. 2017;**89**:16-21

[4] Zhu P, Imai Y, Isoda Y, Shinohara Y, Jia X, Zou G. Composition-dependent thermoelectric properties of PbTe doped with Bi_2Te_3. Journal of Alloys and Compounds. 2006;**420**(1-2):233-236

[5] Christakudi TA, Christakudis GC, Borissova LD. Thermoelectric power of solid solutions $(PbTe)_{1-x}(Bi_2Te_3)_x$ with $0 \leqq x \leqq 0.02$. Physica Status Solidi. 1995;**190**(2):537-544

[6] Gelbstein Y. Simulation of morphological effects on thermoelectric power, thermal and electrical conductivity in multi-phase thermoelectric materials. In: Skipidarov S, Nikitin M, editors. Thermoelectrics for Power Generation. London, UK: IntechOpen; 2016. pp. 286-301

[7] Noda Y, Ohba S, Sato S, Saito Y. Charge distribution and atomic thermal vibration in lead chalcogenide crystals. Acta Crystallographica Section B. 1983;**39**(3):312-317

[8] Shimazaki H, Ozawa T. BiTe, a new mineral from the Tsumo mine, Japan. American Mineralogist. 1978;**63**:1162-1165

[9] Shannon R. Revised effective ionic radii and systematic studies of interatomie distances in halides and chaleogenides. Acta Crystallographica. 1976;**A32**:751-767

[10] Rogacheva E, Vodorez O, Pinegin V, Nashchekina O. Evidence for self-organization processes in PbTe-Bi_2Te_3 semiconductor solid solutions. Journal of Materials Research. 2011;**26**(13):1627-1633

[11] Karpinskii OG, Shelimova LE, Avilov ES, Kretova MA, Zemskov VS. X-ray diffraction study of mixed-layer compounds in the PbTe-Bi_2Te_3 system. Inorganic Materials. 2002;**38**(1):17-24

[12] Gelbstein Y, Dashevsky Z, Dariel MP. High performance *n*-type PbTe-based materials for thermoelectric applications. Physica B: Condensed Matter. 2005;**363**(1-4):196-205

[13] Gelbstein Y. Phase morphology effects on the thermoelectric properties of $Pb_{0.25}Sn_{0.25}Ge_{0.5}Te$. Acta Materialia. 2013;**61**(5):1499-1507

[14] Bergman DJ, Levy O. Thermoelectric properties of a composite medium. Journal of Applied Physics. 1991;**70**(11):6821-6833

[15] Webman I, Jortner J, Cohen MH. Thermoelectric power in inhomogeneous materials. Physical Review B. 1977;**16**(6):2959-2964

[16] Goldsmid H. Bismuth-antimony alloys. Physica Status Solidi. 1970;**7**:7-28

[17] Khalouk K, Chaib C, Gasser J-G. Electrical and thermal conductivities and Seebeck coefficient of liquid copper-bismuth alloys. Philosophical Magazine Letters. 2009;**89**(3):249-262

Section 2

Thermo-Electrical Performance Evaluation of Thermoelectric Devices

Chapter 5

Thermoelectric Generator Using Passive Cooling

Robert Dell, Michael Thomas Petralia, Ashish Pokharel and Runar Unnthorsson

Abstract

This chapter presents an analysis of a point-of-use thermoelectric generator that is patented by one of the authors. The design, implementation and performance of the generator for powering electronic monitoring devices and charging batteries is discussed. This passive generator has no moving parts and relies on ambient air cooling. In one iteration it produces 6.9 W of steady state power using six Laird thermoelectric modules (Laird PB23 Series, HT8, 12) when placed on a 160°C steam pipe with a 30°C ambient environment (ΔT of 130°C). The generator produced 31.2 volts (V) open circuit and 0.89 amperes (A) short circuit. It successfully powered two microcontroller-based security cameras, one with a wireless Local Area Network (LAN) and another with cellular connectivity. In another scenario, the generator produced approximately 6 W with a steam pipe temperature of 140°C and an ambient of 25°C (ΔT of 115°C). This second system powered LED lights, a cellular-interfaced video surveillance system, and monitoring robots, while simultaneously trickle charging batteries. A third installation totally powered a stand-alone 3G web security camera system.

Keywords: thermoelectric generator, passive cooling, low temperature, waste-heat, point of use DC generator, thermoelectric module, steam pipe

1. Introduction

Internal temperature differences in almost all materials cause energetic excitement and the bulk motion of free electrons. This creates an electromotive force called the thermoelectric effect. Metals typically generate a few microvolts per degree Kelvin. Semiconductors can increase this rate by several of orders of magnitude [1, 2].

The thermoelectric effect was reported in the mid-eighteenth century by Franz Ulrich Theodor Aepinus in his study of the mineral tourmaline [3–5]. In 1794, Alessandro Volta reported muscle spasms in a live frog's leg that were caused by two non-ferrous wires immersed in separate glasses of water. The glasses were connected by a partially immersed iron rod that had one end preheated in boiling water [5–7]. The generation of an electromotive force produced by heating the junction between two dissimilar metals came to be known as the Seebeck effect, in honor of Thomas Johann Seebeck, who is credited with the discovery of thermoelectric effects through his experiments in the 1820s [1, 8–13]. Later, in 1834, Jean Charles Athanase Peltier published an article on the inverse effect: temperature anomalies observed near the boundary between two different conductors when a current is passed through them [1, 14]. William Thomson (eventually given the title Baron Kelvin) laid out a thermodynamic basis for thermoelectricity [15] using Joule's law of the electrical generation of heat in a homogeneous metallic conductor.

Continual advances in fundamental physics, materials science, heat transfer, electrical conductivity, and electromotive force-generating capability of thermoelectric materials [2, 16–20] have enabled the development of modern solid-state devices that can directly convert thermal energy to electrical energy. One notable application is in waste heat recovery systems [21, 22]. In 2017, the energy consumption for the United States of America was estimated by the Lawrence Livermore National Laboratory and the U.S. Department of Energy to be 9.77×10^{16} BTU (1.03×10^{20} J). Approximately 68% of this energy was classified as rejected (i.e., wasted) [23]. It is likely that most of the rejected energy was in the form of heat, thermal energy that is transferred between two systems by virtue of a temperature difference [24].

Waste thermal energy is not typically harvested due to its low energy potential. From a thermodynamic perspective, systems with small temperature differences have a low efficiency [25]. Frictional, gravitational, atmospheric, and elastic forces become significant. The resultant basic economics severely limit viable implementations.

Thermoelectric power generation systems can be robust with a relatively simple construction. Despite the low energy potentials of waste heat scenarios and low thermoelectric system efficiencies, they can be an ideal solution in low power applications that have limited electrical grid access and require low maintenance.

2. Background summary

Thermoelectric power generation systems (TEGs) are classical thermodynamic heat engines with no internal moving mechanical parts. An electromotive force is generated through the motion of the Fermi gas of electrons inside conductors and semiconductors (analogous to the molecular gas employed in a steam engine) [26]. A temperature gradient (∇T) induces this electric potential difference or voltage gradient (∇V) in a phenomenon known as either the thermoelectric, Seebeck, or Peltier effect. In general, the bulk material property called the Seebeck coefficient (α), governs thermoelectric behavior. This relationship can be expressed as: $\nabla V = \alpha \nabla T$. The electrochemical phenomenon is an electromotive force induced when two dissimilar metals are connected to form junctions maintained at different temperatures that generates electricity. Individual thermocouples are used to measure temperature when the amount of electricity generated is calibrated to correspond to the temperature. In TEGs, multiple thermocouples called thermoelectric modules (TEMs) are connected in series to generate higher voltages.

TEMs are essentially solid-state devices that convert heat flux into electrical energy. Commercially available TEMs often have more than one hundred thermocouples with pairs of p- and n-type semiconductors connected in series electrically. They are referred to as thermopiles. The thermopiles are typically sandwiched between two thin ceramic plates providing a platform to make both the electrical junctions and thermal interfaces. Two wires at the ends of the thermopile serve as the electrical connections for the TEM. If the opposing sides of the ceramic plates are maintained at different temperatures, direct current (DC) will be generated when the wires are connected to create a circuit. A simple TEG therefore consists of TEMs, a circuit load, a high-temperature heat transfer system (hot block), and a low-temperature heat transfer system (cold block). The two heat transfer systems are required to maintain the temperature difference between both sides of the TEM.

When electricity is added to a TEM, heat is moved from one ceramic surface to the other, which renders the TEM in its traditional role as a solid-state heat pump.

3. Generator system description

The authors have designed and patented a TEG that relies solely on passive cooling; i.e., natural conduction [27]. It can be easily mounted onto an existing pipe heat source.

Figure 1 shows 3D CAD models of the generator. It consists of a pair of mirrored assemblies. Each assembly has a hot block transition that doubles as a surface geometry adapter for three thermoelectric modules that are wired in series, and a cold block with a finned heat pipe [28].

The hot block transfers the heat from the pipe to the TEMs and the cold block removes the heat. The resultant temperature difference between the hot and cold blocks produces thermoelectric power when TEMs are sandwiched in between them. Multiple TEMs in a thermoelectric generator can boost the voltage when wired in series and the amperage when wired in parallel. The generator has a fast reaction; e.g., changes in airflow are almost immediately manifested in measurable voltage and amperage fluctuations.

The remainder of this section will describe in detail the generator's components, the generator's assembly and the setup of experiments carried out to test the design.

3.1 Heat source

Steel steam and hot water pipes in Iceland and New York City were used as test beds [28]. The steam and hot water pipes have internal temperatures between 85 and 160°C. The nominal dimensions of the pipe diameters used by the authors range from 1 to 3.5 in. Exterior and interior locations are used with preexisting and new dedicated pipes. In Iceland direct borehole lines and municipal geothermal hot water heat sources are used. In New York City, Consolidated Edison municipal steam lines were used for three early test configurations. A dedicated steam experimental test bed was developed at the Center for Innovation and Applied Technology at The Cooper Union for the Advancement of Science and Art.

The experimental test bed as shown on the left in **Figure 2** has an ESG Corporation 240-volt 3 phase SPEEDYELECTRIC 15A-2 electrode steam generator with a BFCR-404 automatic water feed pump and condensate return system. It has a

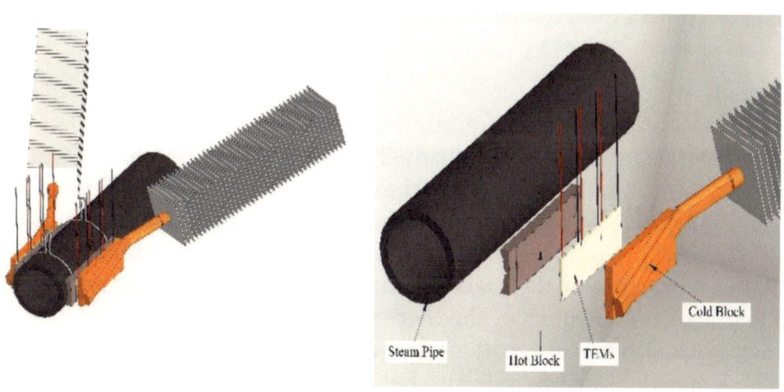

Figure 1.
3D CAD models of the thermoelectric generator assembled as a mirrored unit of two generators (left) and a schematic (right).

Figure 2.
The test beds at The Cooper Union in New York City (left) and the Agricultural University of Iceland in Hveragerdi (right).

Parameter	Value
Internal pipe temperatures	85–160°C
Steam pipe thickness	0.19 in
Steam pipe outer diameter	1–3.5 in
Operating pressure	80–100 psig
Reservoir volume	30 gal

Table 1.
Test bed heat source parameters.

30-gallon water reservoir and the condensate is bled manually. The system operating pressure was set at a maximum pressure of 100 psig and an 80 psig minimum when the unit automatically powered the electrodes until a maximum pressure was again obtained. The steel pipes have 2-in diameter pipes with a 1-in top pipe in a closed loop configuration. The thermoelectric generator is mounted on the bottom 2-in pipe.

The parameters for the heat source in the test bed are listed in **Table 1**. Shown on the right of **Figure 2**, at the Agricultural University of Iceland in Hveragerdi, the thermoelectric generators were attached outdoors to a geothermal steam borehole line. The generators were placed in adjoining protective sheds and attached to an old galvanized pipe with a temperature of between 100 and 140°C.

3.2 Hot block

The hot block creates a thermal channel to the TEMs. Although the hot block material should be chosen to maximize thermal conductivity, non-ferrous metals were not initially selected to prevent potential galvanic reactions between the hot block and the steel pipes. Steel hot blocks would also have the additional advantage of having the same rates of thermal contraction and expansion as the pipes, which helps to maintain the integrity of the thermal grease interfaces.

The hot block also has to accommodate a round steam pipe and a flat TEM, so the hot block is curved on one side and flat on the opposite side, seen in **Figure 3**.

Parallel lateral grooves were milled on the curved side of the hot block. This helped mitigate any differences in radius between the hot block and the steam pipe while also minimizing potential hot block warp. Warping of the hot block would

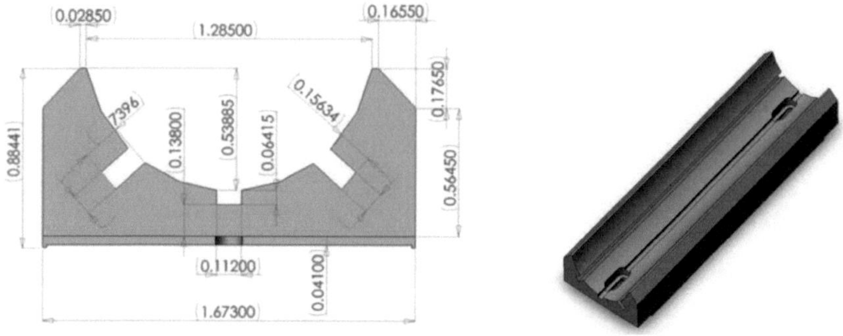

Figure 3.
Cross section with dimensions in inches (left) and 3D rendering of the hot block (right).

Parameter	Value
Average block thickness	0.68 in
Block/module contact area	6.8 in^2
Pipe contact area	7.0 in^2
Number of grooves	3

Table 2.
Hot block design parameters.

degrade the thermal integrity of the interfaces, and possibly damage the TEMs. A third function of the grooves is to expel excess thermal grease.

On the flat side of the authors' hot block, recesses were milled leaving slightly raised lateral flat rails on the edges. This forms a channel for the TEMs, which are slid into a precise controlled position with thermal grease between the surfaces. In traditional TEM installations without the recess and rails, the thermal grease inhibits precise positioning due to its lubricity.

The design parameters for the hot block are listed in **Table 2**.

3.3 Thermoelectric modules

The TEM used are tolerance lapped. They vary no more than 0.005 in. The surface area of contact with the hot and the cold blocks is approximately 1.5 in^2. Each mirrored side of the TEG has three TEMs in contact with the hot block, for a total of six TEMs wired in series for power production. ThermaTEC modules (Laird PB23 Series, HT8, 12) were designed to function in the higher temperature ranges of steam systems. Earlier versions of the TEG used Melcor model HT-4-12-40 modules, which provided lower amperage and higher voltage. Both modules meet or exceed current U.S. military specifications.

The design parameters for the module array are listed in **Table 3**.

3.4 Cold block

The cold block system was fabricated by Noren Products. The system consists of two mirrored heat pipes that are splayed back along the steam pipe. This increases the available space near the steam pipe. The geometry is important for safety concerns, and it enables installation in a one-foot radius envelope around the steam pipe. An added advantage is that the natural convection current around the steam pipe accelerates the cooling.

Parameter	Value
Number of modules	3
Seebeck coefficient	0.050 V/K
Thermal conductance	0.54 W/K
Module side dimension	1.5 in

Table 3.
Thermoelectric module array design parameters.

Parameter	Value
Fin thickness	0.02 in (aluminum) 0.03 in (copper)
Fin length	3–4 in
Fin spacing	0.25 in
Heat pipe outer diameter	0.66 in
Number of fins	41

Table 4.
Cold block design parameters for initial unit.

The main function of the cold block system is to provide a thermal channel between the cold plate of the thermoelectric module array and the ambient environment. The authors' thermoelectric generator is cooled passively by the ambient air. Cold blocks and their usual array of fins are sometimes referred to as "heat sinks". In a larger thermal sense, they are only conduits to the heat sink, which is actually the immediate environment. This can be an outdoor setting or a confined space, such as a steam room, where an additional thermal analysis is needed for an accurate design assessment. The authors' cold block system consists of a thick milled copper plate, in direct contact to the TEM array. It is soldered to a copper heat pipe with attached square aluminum fins.

The design parameters for the cold block are listed in **Table 4**.

3.4.1 Heat pipe

Heat pipes are essentially heat siphons that rely on fluid phase change to greatly increase their heat transfer ability. They are different from traditional heat siphons, such as single pipe steam heating systems, because they have a porous layer next to the pipe's surface that accelerates the liquid flow by using capillary action caused by the increased surface area. There is often a smaller concentric internal pipe that serves as a dedicated return line.

Heat pipe transmission rates in terms of watts per unit area are more than 50 times greater than copper [29]. The internal fluid flow of the heat pipe relies in part on gravity, and it performs poorly if placed horizontally. A mild angle (e.g., 15 degrees from horizontal) is sufficient for the heat pipe chosen.

Heat pipes are also lightweight, weighing about one-fifth of traditional pipes for the same cooling effectiveness. A traditional heat pipe may create substantial stress on the TEMs due to the cantilevered weight. Heat pipes are also somewhat flexible.

3.4.2 Fins

Rectangular fins are mounted onto the heat pipe with a heat shrink fit. The system chosen relies primarily on natural convection, and a restriction on the fins is that they must be nearly vertical to allow the flow to accelerate vertically between them. The spacing must also be sufficiently large that there is minimal interference between the thermal boundary layers of adjacent fins.

There are forty-one fins on each heat pipe assembly. They were 3 and 4 in^2. The output voltage is predicted to increase with increasing numbers of fins due to the increase in surface area, but there are diminishing returns. Doubling the number of fins to 82 (and lengthening the heat pipe to maintain the proper spacing) results in a modest 22% increase in output volt predicted (18.1–22.1 V), as can be seen in **Figure 4**. The longer heat pipe would also double the weight and be more unwieldy than the design chosen. The loss of robustness was determined to be more important than the projected power increase. Improvements in performance were achieved by increasing the fin dimension *L*. With 4-in fins, the output voltage was increased by 16%, the modification representing an 87% increase in surface area.

As expected, there is a slightly higher voltage for copper compared with aluminum fins. Copper has nearly double the thermal conductivity of aluminum, and the copper fins have a larger thickness in the prototype (0.02 in for aluminum and 0.03 in for copper). However, the difference is small because the fins are effectively thermally short. That is, the characteristic fin lengths are 13 and 7.5 in for copper and aluminum, respectively, whereas the fin dimension was only 3 in, less than half the characteristic lengths, so the temperature drop along the fin surface is small.

The natural convection coefficient increases with the decreasing vertical dimensions of the fins. Therefore, another means of improving performance (for thermally short fins) would be to use rectangular, rather than square, fins. For example, under these conditions, the convection coefficient is 5.9 W/m^2 °C for 3-in fins and 7.2 W/m^2 °C for 2-in fins. The predicted output voltage increases from 18.4 to 19.5 V if fins of 2 in (vertical) by 4.5 in are used, a 6.0% increase without increasing the fin area.

3.5 Thermal grease

Thermal greases can be employed to increase the thermal conductivity of the thermal interfaces. Traditional thermal greases are not recommended by the manufacturers for the applications described herein as they quickly dry at steam temperatures. A specifically modified version of the Ambrosia HT high-temperature

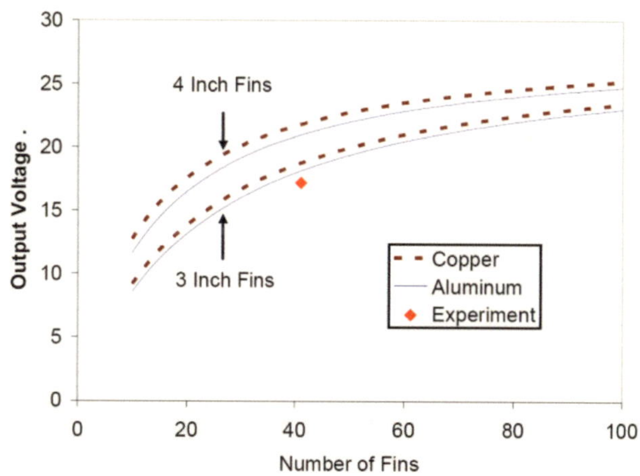

Figure 4.
System open-circuit voltage as a function of the number of fins for copper and aluminum fins of 3-in (used in the prototype) and 4-in dimension. Experimental point and theoretical predictions are obtained from a pair of thermoelectric assemblies.

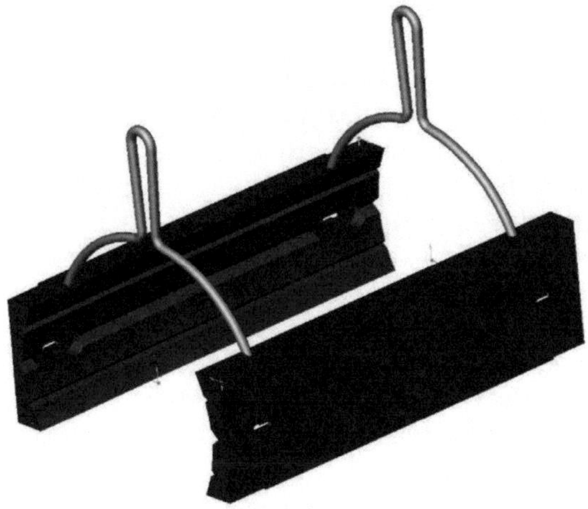

Figure 5.
CAD rendering of a pair of hot blocks connected by 2 cradles which act as springs and apply a clamping force.

thermal grease from Artic Silver was employed. It contains nanoparticles that settle into any material voids over a period of approximately 100 h, thereby potentially increasing the system power. Later installations used Timtronics Red Ice thermal grease.

The thermal grease also doubles as a galvanic corrosion barrier between the dissimilar metals.

3.6 Clamping system

Clamps provide the pressure needed for basic attachment and to maintain the integrity of the thermal interfaces. A double wire spring connects the mirrored hot blocks in a cradle configuration for alignment and initial positioning purposes. Additionally, this cradle system serves as gripping points during installation and a connection point for the wire harnesses. A CAD rendering of the wire spring clamp and the hot blocks is shown in **Figure 5**.

3.7 Electrical energy storage

Batteries are traditionally used to store electrical energy for later use. In thermoelectric applications, a battery system can provide peak power. It also eliminates the inherent electrical noise that can interfere with the solid-state telemetry systems. Rechargeable lead-acid, NiCad, lithium-ion, or other battery storage can be used.

4. Theoretical considerations

4.1 Thermoelectric power

The theoretical power generated can be derived by modeling a TEM as a voltage source (V_o) with an internal resistance (r_i) and a circuit load resistance (r_L). If the module is short circuited (closed, with zero load resistance) the maximum electric current (I_c) is developed from the open circuit voltage with only the TEM's internal resistance. The short circuit current is given by Ohm's law: $I_c = V_o/r_i$. When an electrical load is placed across the module, the circuit current is decreased due to the additional series resistance. The loaded circuit current is given by $I_L = V_o/(r_i + r_L)$. Here Ohm's law can be employed again, this time to determine the voltage across the load:

$$V_L = \left(\frac{r_L}{r_i + r_L}\right)V_o. \quad (1)$$

Since power is the energy output over time, voltage is the potential energy per electric charge, and current is the electric charge flow over time, the power available at the load of an electric circuit is given by $P_L = V_L I_L$. Thus, the power generated by a TEM can be expressed as

$$P_L = \left[\frac{r_L}{(r_i + r_L)^2}\right]V_o^2. \quad (2)$$

By solving $dP_L/dr_L = 0$, the maximum power is determined to occur when $r_L = r_i$, which is the classic case of impedance matching. The maximum power achievable can thus be written in terms of the open circuit voltage and the short circuit current:

$$P_{L,\max} = \frac{V_o I_c}{4}. \quad (3)$$

TEMs have multiple thermocouples connected in series and thermally in parallel. The plates on either side of the thermopile are at T_h and T_c. The power generated by the TEM can be expressed as

$$P_L = \left[\frac{r_L}{(r_i + r_L)^2}\right]\alpha^2(T_h - T_c)^2. \quad (4)$$

4.2 Thermal resistance modeling

Thermal resistance is analogous to electrical resistance. The total resistance of a thermal circuit is therefore the cumulative total of the individual thermal resistors [30]. Overall system efficiency can be increased by lowering these thermal resistances. TEMS are placed between the hot and the cold blocks. Any other thermal connection would tend to reduce the change in temperature (ΔT) across the TEMs, which degrades the generated power.

The cold block's copper plate and the attached heat pipe are very thermally conductive when compared to the steel hot block and the TEM array. It can be regarded as de minimus, and therefore it is not included in the simplified analytical model. The thermal resistance associated with the connection of the heat pipe to the aluminum fins is also de minimus due to the minimal area of interface.

The base of each fin's contact with the heat pipe is maintained at the temperature of the cold side of the TEM.

4.2.1 Hot block

The heat source is the pipe. The heat flows from the pipe fluid to the inner surface of the steam pipe, the pipe wall, and a thermal grease interface. This connects to the hot block, another thermal grease interface and then to the hot side of the TEM.

The hot block system, R_H, consists mainly of three resistances in series:

$$R_H = R_{steam} + R_{pipe}(+R_{contact\ 1}) + R_{hotblock}(+R_{contact\ 2}). \tag{5}$$

The two contact resistances (thermal grease) shown in parentheses, between the steam pipe and the hot block, and between the hot block and the TEM, can be considered negligible if good thermal contact is maintained using an efficient thermal grease.

The thermal resistance between the steam and the pipe inner wall, R_{steam}, is a convective resistance and is approximated by

$$R_{steam} = \frac{1}{h_s A_s} \tag{6}$$

where A_s is the effective surface area and h_s is the convective coefficient of the condensing steam on the inside of the pipe wall. A_s will be somewhat larger than the contact area of the hot block due to three-dimensional effects.

The thermal resistance across the pipe wall is a conductive resistance and is approximated by:

$$R_{pipe} = \frac{t_p}{k_p A_p} \tag{7}$$

where A_p is the effective conductive area, k_p is the thermal conductivity, and t_p is the thickness of the pipe wall. A_p will again be somewhat larger than the contact area of the hot block due to three-dimensional effects.

The thermal resistance across the hot block is also a conductive resistance that is approximated by:

$$R_{hotblock} = \frac{t_{hb}}{k_{hb} A_{hb}} \tag{8}$$

where A_{hb} is taken to be the average area of the two contact surfaces of the hot block, k_{hb} is the thermal conductivity of the hot block, and t_{hb} is the average thickness of the block.

The total thermal resistance of the hot block system is thus approximated by

$$R_H = \frac{1}{h_s A_s} + \frac{t_p}{k_p A_p} + \frac{t_{hb}}{k_{hb} A_{hb}} \tag{9}$$

Note that approximately half of the thermal resistance is due to the hot block itself and the other half is due to a combination of condensation and conduction through the pipe wall. That is, the thermal resistance added with the hot block is comparable to the inherent thermal resistance of the existing pipe.

4.2.2 TEMs

The TEMs are thermally connected in parallel, so the thermal resistance for three modules is:

$$\frac{1}{R_{TEM}} = \frac{1}{R_{m,1}} + \frac{1}{R_{m,2}} + \frac{1}{R_{m,3}} \tag{10}$$

where $R_{m,i}$ is the thermal resistance of the ith TEM. Assuming all the TEMs have the same thermal resistance, R_m, this simplifies to:

$$R_{TEM} = \frac{R_m}{3}. \tag{11}$$

The thermal conductance of each module is taken to be 0.5 W/°C at 100°C. (The thermal conductance the inverse of the thermal resistance.) The thermal resistance of the three-TEM array is:

$$R_{TEM} = \frac{\frac{1}{0.5}{}^{\circ}\mathrm{C/W}}{3} = 0.67^{\circ}\mathrm{C/W}. \tag{12}$$

4.2.3 Cold block

A simplified model of the cold block can be represented by a system of essentially parallel resistances. The cold block resistance, R_C, for N fins becomes:

$$\frac{1}{R_C} = \frac{1}{R_1} + \frac{1}{R_2} + \frac{1}{R_3} + \cdots + \frac{1}{R_{N-1}} + \frac{1}{R_N}. \tag{13}$$

Assuming all the thermal resistances of the fins are equal ($R_1 = R_2 = R_3 = \cdots = R_{N-1} = R_N = R_{fin}$), the system reduces to:

$$R_C = \frac{R_{fin}}{N}. \tag{14}$$

An estimate of the thermal resistance of the individual fins can be modeled as two thermal resistors in series:

$$R_{fin} = R_k + R_h. \tag{15}$$

where R_k is the thermal resistance associated thermal conduction within the fin material from the base toward the outer perimeter, and R_h is the thermal resistance associated with heat being transferred from the fin to the ambient air.

5. Results

The first test bed at The Cooper Union in New York City had a 3 in diameter steam pipe and ambient air temperatures of approximately 350°F (177°C) and 122°F (50°C), respectively. The ΔT was approximately 120°C. For the initial prototype, the first assembly had copper fins and the second had aluminum fins. The prototype had a voltage of 17.2 V, compared with the theoretical prediction of 18.4 V, which was the average of the 18.1 V predicted for aluminum and 18.8 V predicted for copper. These results demonstrate the output open circuit voltage, not how the system responds to an electrical load (with current being drawn). Therefore, the system voltage and current were also measured when the system was used to drive two different light bulbs. The closed circuit current (zero voltage) was found to be 0.70 A. A linear relationship was found experimentally to be

$$V = -24.4I + 17.2 \tag{16}$$

With the first assembly constructed with copper fins and the second with aluminum fins, the direct current system produced 17.2 V (open circuit) and 0.7 A (closed circuit). A theoretical methodology for predicting the output electrical power from an array of thermoelectric modules is developed and applied to this system.

As can be seen in **Figure 4**, the theoretical system output and the output was found to be extremely stable overtime. The theoretical model can therefore be used to predict the effect of changes in design parameters or operating conditions.

The output was found to be extremely stable over time.

The third assembly was constructed and tested to determine whether a heat pipe was necessary. The 16 mm heat pipe was replaced with a 38 mm solid copper rod manufactured by Noren Products and the fin size was adjusted so the system would yield comparable results to the first two assemblies. The theoretical model was extended to include the thermal resistance of this rod (which was assumed to be zero for the heat pipe). The solid copper rod weighs approximately 6 kg, as compared to 1.5 kg for the heat pipe. The solid copper assembly demanded considerable extra fabrication costs due to extensive welding, soldering, and extra machining.

The fourth design iteration increased the fin size from 3 to 4 in. The output increased from 8.18 to 11.54 V (a 40% increase). The experimental output of this fourth assembly was 14% higher than theoretically predicted. The third and fourth assemblies can be compared to note the effect of the heat pipe over the solid copper rod because both had 4-in fins.

The fifth and sixth assemblies tested were fit to a 1-in steam pipe at a Consolidated Edison facility in New York City. They were otherwise designed identically to the fourth assembly, with 4-in fins and a heat pipe. The fifth assembly contained three TEMs and had a measured open circuit voltage of 15.9 V, while the sixth only contained 2 TEMs and had a measured open circuit voltage of 12.1 V. These two assemblies provided uninterrupted power for at least 2 years.

Three assemblies were eventually moved to an outdoor location in Hveragerði, Iceland at the University of Iceland, where they were fit onto old galvanized pipes with surface temperatures of approximately 140°C from a geothermal steam borehole line. The steam pipe has been a reliable power source for almost 10 years. The total voltage of the three units in series after 9 months of exterior exposure was 39.2 V (open circuit) and the amperage was 0.64 (short circuit). The output in parallel was 13.4 V (open circuit) and 1.81 A (short circuit). The power produced was approximately 6W.

The TEMs were replaced and the new Timtronics Red Ice thermal grease was applied. In the winter of 2019 with a ΔT of 140°C, the open circuit voltage and short circuit current of the three assemblies were each measured independently in steady state, as is shown in **Table 5**. When the assemblies are wired in series, the voltages sum to 31.6 V and the power is approximately 10.7 W. The power generated by the TEGs was 257 Wh per day. **Figure 6** shows the total energy drawn from the TEGs. The energy drawn was 61 Wh per day. The entire 3G web camera system and its controls consume only 25% of the total power generated. In 2019 the test bed at the Center for Innovation and Applied Technology at The Cooper Union had steady-state open circuit voltage and short circuit current measuring 19.32 V and 0.96 A. The experimental data is represented in **Figure 7**. The maximum power

Assembly name	Voltage	Current
Generator 1	10.17 V	1.35 A
Generator 2	10.63 V	1.35 A
Generator 3	10.79 V	1.36 A

Table 5.
January 2019 Hveragerdi steady state voltage and amperage readings.

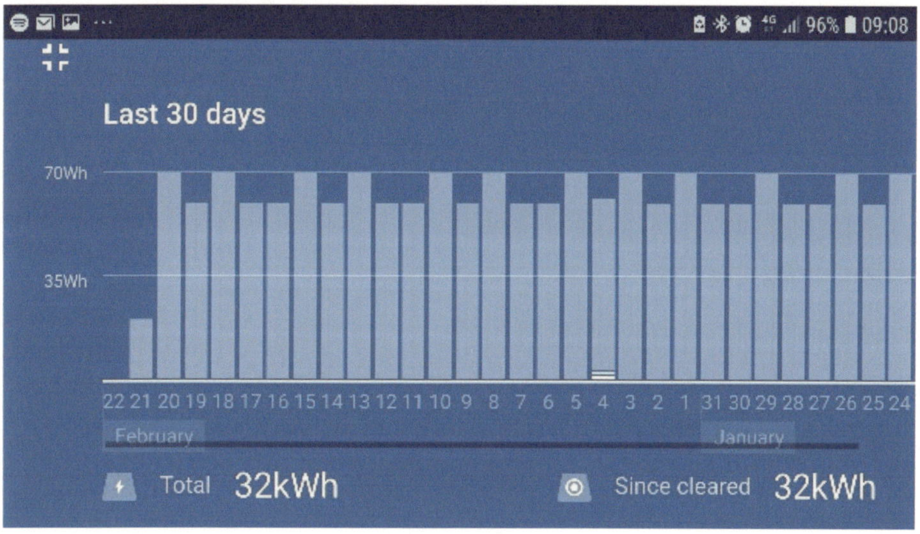

Figure 6.
January 24 to February 21, 2019 3G system power consumption. The February 21 reading is less due to the time of day the readings were taken.

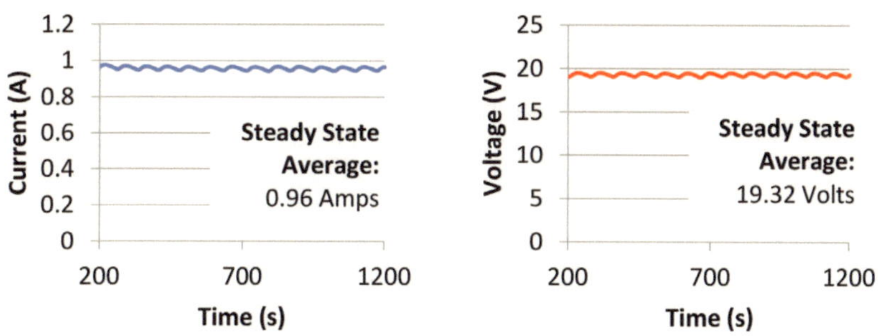

Figure 7.
Experimental electrical power generation data from the New York City test bed recorded in January 2019. There is a steady-state average short-circuit current of 0.96 A (left) and a steady-state average open-circuit voltage of 19.32 V (right).

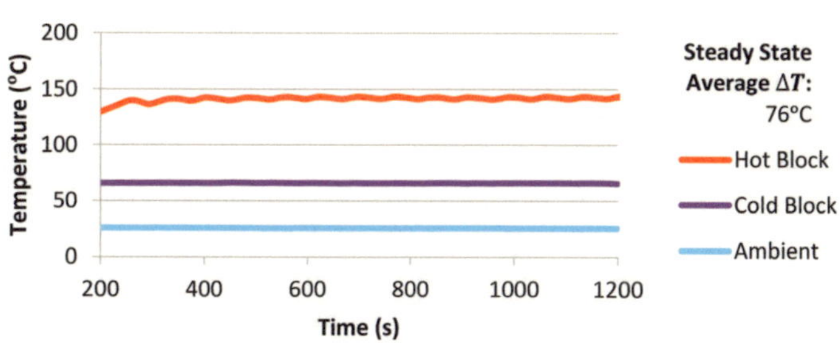

Figure 8.
Experimental temperature difference (ΔT) data from the New York City test bed recorded in January 2019. There is a steady-state temperature difference of 76°C.

available, then, is 4.6 W. The hot block, cold block, and ambient temperatures were 142, 66, and 27°C, respectively. The experimental data is represented in **Figure 8**.

Figure 9 illustrates the linear voltage-amperage of The Cooper Union thermoelectric generator. A Fluke 289 graphical multimeter using Flukeview software was used to record all of the measurements.

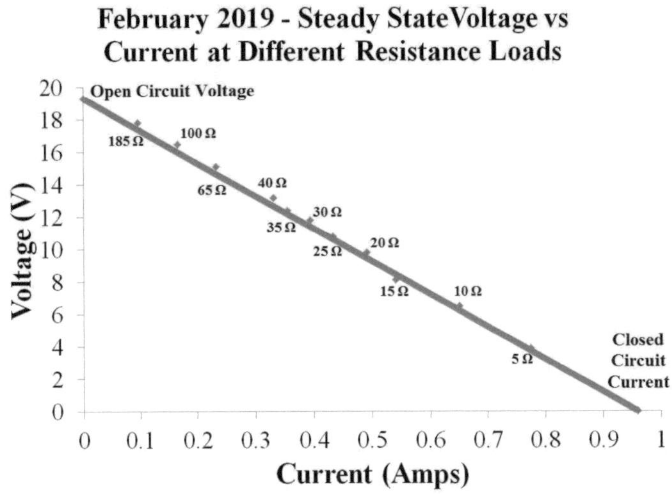

Figure 9.
Steady state voltage versus amperage linear relationship data from the New York City test bed recorded in February 2019.

6. Applications

A primary application of TEGs using passive cooling is in remote locations where normal powering methods for monitoring systems are problematic due to distance from the electrical grid. Where there are geothermal bore holes and steam pipes, the TEG described is a direct application and will function well, especially when ambient temperatures are low. In contrast, solar power options are weather dependent, limited during the winter months, and colder temperatures are detrimental to their stand-alone batteries.

Developments in mobile and handheld electronics have enabled low-power designs with new connectivity and energy-saving scenarios through incorporation of integrated circuits and microcontroller units. As the processing power of computers increased, the need for transferring and sharing information became stronger. Monitoring systems, which used to be run as stand-alone systems, are now networked with other systems. In many cases, these systems are also control systems. Information sharing provides a clearer overall picture of the process and allows for more intelligent control.

The need for real time data or simplified data collection from multiple sensors has driven the development of wired and wireless sensor solutions. In the case of wired connections, the need for an external power supply is often eliminated by enabling the sensors to function as parasites and draw their power from the data line. Power for running wireless systems is normally obtained from either batteries or internal power supplies. The advances in both low-power electronics and ambient energy harvesting have led to significant industrial and academic research on wire- and battery-free sensors and devices. Wireless sensors powered by ambient energy can be found in transportation, building automation, industry, and other applications. This technology eliminates the need to buy and replace batteries, and units can be left unattended in hazardous or difficult-to-get-to places.

The authors have developed an intensive open-field heat agriculture system using geothermal steam and steam condensate that extends growing seasons and enables the cultivation of out-of-region, non-native crops [31, 32]. It has an underground piping system in Iceland that is similar to the method for heating sidewalks and how waste heat from combined heat and power (CHP) systems warm green roofs [33]. The TEG described herein has been used to power automated irrigation systems [34], robotic remote monitoring systems [35], and a series of web-accessible security cameras working on both LAN and 3G wireless networks in New York City and Iceland, respectively [36–39].

7. Conclusions

The passive TEG described has no moving parts and relies on ambient air cooling. It can maintain a temperature difference (ΔT) across a series of TEMs of 76°C. In one iteration, it produces 6.9 W of steady state power with 31.2 V open circuit and 0.89 A short circuit. Successful applications include powering two microcontroller-based security cameras, one with a wireless Local Area Network (LAN) and another with cellular connectivity, as well as LED lights, a cellular-interfaced video surveillance system, and monitoring robots, while simultaneously trickle charging batteries. The most recent application was to power a complete 3G wireless security network. Future work includes optimizing the cold block to further increase the ΔT across the TEMs.

Acknowledgements

The authors acknowledge the support extended by the following organizations, institutions and corporations: The Cooper Union for the Advancement of Science and Art, the Center for Innovation and Applied Technology, the Agricultural University of Iceland, the University of Iceland, Arvirkinn ehf, Timtronics, the City of Hveragerdi, and Keilir Institute of Technology. Special thanks to: Gudmundur Gislason, Ruth Nerken, Mark Epstein, Barry Shoop, Richard Stock, Anita Raja, Melody Baglione, George Sidebotham, Chih S. Wei, Gudridur Helgadottir, Aldis Hafsteinsdottir, Elias Oskarsson, Borkur Hrafnkelsson, Mar Gudmundsson, Fridrik Brekkan, Stefan Sigurdsson, Gísli Páll Pálsson and Aladino Melendez. The authors acknowledge the contributions of the Center for Innovation and Applied Technology Research Assistants: Daniel Abes, Jeahoung Hong, Seung Won Na, Jabin Pu, Christopher E, Jing Jin, Sanjev Menon, Di Yi Liu, Daniel Feyman, Alinur Rahim, Justin Jose, James Ngai, Hou Chong Chan, Issei Abraham Yamada, Wei Yan Tin, Chengyin Jiang, Yueyue Li, TaeKoung Lee, Harrison Milne, Romaniya Voloshchuk, Monica Chen, Jordan Selig, and Matthew Cavallaro.

Conflict of interest

The authors declare no conflict of interest. The founding sponsors had no role in the design of the study; in the collection, analyses, or interpretation of data; in the writing of the manuscript, and in the decision to publish the results.

Nomenclature

Δ	change
∇	gradient
α	Seebeck coefficient
A_{hb}	hot block effective surface area
A_s	steam effective surface area
A_p	pipe effective surface area

h_s	steam convective coefficient
I	current
I_c	short circuit current
I_L	load current
k_{hb}	hot block thermal conductivity
k_p	pipe thermal conductivity
n_m	number of TEMs in the array
N	number of fins
P_L	load power
r_i	internal resistance
r_L	load resistance
R_C	cold block thermal resistance
$R_{contact}$	contact thermal resistance
R_{fin}	fin thermal resistance
R_h	fin convective thermal resistance
R_H	hot block system thermal resistance
$R_{hotblock}$	hot block thermal resistance
R_k	fin conductive thermal resistance
R_m	TEM thermal resistance
R_{TEM}	TEM array thermal resistance
R_{steam}	steam thermal resistance
R_{pipe}	pipe thermal resistance
t_{hb}	hot block average thickness
t_p	pipe wall thickness
T	temperature
T_c	hot side temperature
T_h	cold side temperature
V	voltage
V_o	open circuit voltage
V_L	load voltage

Author details

Robert Dell[1*], Michael Thomas Petralia[1], Ashish Pokharel[1] and Runar Unnthorsson[2]

1 Center for Innovation and Applied Technology, Mechanical Engineering, The Cooper Union, New York, NY, USA

2 Faculty of Industrial Engineering, Mechanical Engineering, and Computer Science, University of Iceland, Reykjavik, Iceland

*Address all correspondence to: rdell@cooper.edu

References

[1] Ioffe A. Semiconductor Thermoelements and Thermoelectric Cooling. London: Infosearch Limited; 1957. 184 p

[2] Pollock D. Thermocouples: Theory and Properties. Boca Raton: CRC Press, Inc.; 1991. 336 p LCCN: 91-24377

[3] Aepinus F. Tentamen Theoriae Electricitatis et Magnetism. Petropoli: Typis Academiae Scientiarum; 1759. 390 p

[4] Home R. Aepinus, the tourmaline crystal, and the theory of electricity and magnetism. Isis. 1976;**1**:21-30. DOI: 10.1086/351542

[5] Goupil C et al. Thermodynamics and thermoelectricity. In: Goupil C, editor. Continuum Theory and Modeling of Thermoelectric Elements. Weinheim: Wiley-VCH; 2015. pp. 1-73. DOI: 10.1002/9783527338405.ch1

[6] Volta A. Nuova memoria sull'elettricità animale, divisa in tre lettere, dirette al signor abate Anton Maria Vassalli Lettera prima. Annali di Chimica e Storia Naturale. 1794;**5**: 132-144

[7] Wikipedia, Thermoelectric effect: Note 1 [Internet]. 2019. Available from: https://en.wikipedia.org/wiki/Thermoelectric_effect [Accessed 2019-01-09]

[8] Seebeck T. Über den Magnetismus der Galvanischen Kette. Berlin: der Koniglichen Akademie der Wissenschaft; 1822. 58 p

[9] Oersted M. Nouvelles expériences de M. Seebeck sur les actions électro-magnetiques. Annales de Chimie et de Physique. 1823;**22**:199-201

[10] Oersted M. Notiz von neuen electrisch-magnetischen Versuchen des Herrn Seebeck in Berlin. Annalen der Physik. 1823;**73**:430-432

[11] Seebeck T. Magnetische polarisation der metalle und erze durch temperatur-differenz. Abhandlungen der Königlichen Akademie der Wissenschaften zu Berlin. 1825:265-373. https://books.google.com/books?id=E27NAAAAMAAJ&pg=PA265#v=onepage&q&f=false

[12] Seebeck T. Ueber die magnetische polarisation der metalle und erze durch temperatur-differenz. Annalen der Physik und Chemie. 1826;**6**:1-20, 133-160, 253-286

[13] Goldsmid H. Introduction to Thermoelectricity. 2nd ed. Berlin: Springer-Verlag; 2016. 278 p. DOI: 10.1007/9783662492567

[14] Peltier J. Nouvelles expériences sur la caloricité des courants électrique. Annales de Chimie et de Physique. 1834; **56**:371-386

[15] Thomson W. On a mechanical theory of thermo-electric currents. Proceedings of the Royal Society of Edinburgh. 1851;**42**:91-98

[16] Heikes R, Ure R. Thermoelectricity: Science and Engineering. New York: Interscience Publishers, Inc.; 1961. LCCN: 61-9650

[17] Harman T, Honig J. Thermoelectric and Thermomagnetic Effects and Applications. New York: McGraw-Hill, Inc.; 1967 LCCN: 66-20717

[18] Tian Z, Lee S, Chen G. Comprehensive review of heat transfer in thermoelectric materials and devices. Annual Review of Heat Transfer. 2014;**17**:425-483. DOI: 10.1615/AnnualRevHeatTransfer.2014006932

[19] Zhang X, Zhao L. Thermoelectric materials: Energy conversion between

heat and electricity. Journal of Materiomics. 2015;**1**:92-105. DOI: 10.1016/j.jmat.2015.01.001

[20] Ge Z et al. Low-cost, abundant binary sulfides as promising thermoelectric materials. Matrials Today. 2016;**4**:227-239. DOI: 10.1016/j.mattod.2015.10.004

[21] LeBlanc S. Thermoelectric generators: Linking material properties and systems engineering for waste heat recovery applications. Sustainable Materials and Technologies. 2014;**1-2**: 26-35. DOI: 10.1016/j.susmat.2014.11.002

[22] Paul D. Thermoelectric energy harvesting. In: Fagas G, Gammaitoni L, Paul D, editors. ICT–Energy–Concepts Towards Zero: Power Information and Communications Technology. London: Intech; 2014. 49-78 p. DOI: 10.5772/55410

[23] Lawrence Livermore National Laboratory, Energy Flow Charts: Estimated U.S. Energy Consumption in 2017: 97.7 Quads [Internet]. 2019. Available from: https://flowcharts.llnl.gov/content/assets/docs/2017_United-States_Energy.pdf [Accessed: 2019-01-09]

[24] Lee H. Thermal Design: Heat Sinks, Thermoelectrics, Heat Pipes, Compact Heat Exchangers, and Solar Cells. Hoboken: John Wiley & Sons, Inc.; 2010. 2 p. DOI: 10.1002/9780470949979

[25] Çengel Y, Boles M. Thermodynamics: An Engineering Approach. 4th ed. New York: McGraw-Hill; 2002. 275. p 10:0072383321

[26] Goupil C et al. Thermodynamics of thermoelectric phenomena and applications. Entropy. 2011;**13**: 1481-1517. DOI: 10.3390/e13081481

[27] Dell R, Wei S, Sidebotham G. Thermoelectric Power Generation Device. European Patent No. EP 2,095,440 B1, U.S. Patent No. US 8,829,326 B2, and 9,590,160, Canadian Patent No. CA 2,671,995, 7 March 2017

[28] Dell R et al. A thermoelectric-based point of use power generator for steam pipes. Geothermal Resources Council Transactions. 2011;**35**:1237-1241

[29] Dell R et al. Thermoelectric-based power generator for powering microcontroller based security camera. In: Proceedings of the ASME 2012 International Mechanical Engineering Congress and Exposition (IMECE2012); 9–15 November 2012. Houston, USA: IMECE; 2012. pp. 635-642. DOI: 10.1115/IMECE2012-89611

[30] Sidebotham G. Heat Transfer Modeling. Cham: Springer; 2015. viii p. DOI: 10.1007/9783319145143

[31] Dell R et al. Repurposing waste steam and hot water to accelerate plant growth in heated green roofs. In: Proceedings of the ASME 2013 International Mechanical Engineering Congress and Exposition (IMECE2013); 15-21 November 2013. San Diego, USA: IMECE; 2013. DOI: 10.1115/IMECE2013-65200

[32] Dell R et al. Waste geothermal hot water for enhanced outdoor agricultural production. In: Proceedings of the ASME 2013 Power Conference (POWER2013); 29 July–August 1 2013. Boston, USA: POWER; 2013. DOI: 10.1115/POWER2013-98172

[33] Dell R et al. Geothermal heat in agriculture: Preliminary results of an energy intensive system in Iceland. Geothermal Resources Council Transactions. 2011;**35**:115-121

[34] Foley W et al. Point of use thermoelectric powered automated irrigation system for an intensive shallow bottom heat system using waste geothermal hot water and steam

condensate in Iceland. Geothermal Resources Council Transactions. 2015; **39**:117-123

[35] Dell R et al. A thermoelectric powered quadruped robotic system for remote monitoring of geothermal open field heated gardens in Iceland. Geothermal Resources Council Transactions. 2016;**40**:173-180

[36] Dell R et al. Web accessible security camera system independently powered by a point of use thermoelectric generator using geothermal pipes as a heat source. Geothermal Resources Council Transactions. 2014;**38**:729-733

[37] Dell R et al. A web-accessible robotics monitoring system powered by a thermoelectric generator connected to a battery. In: Proceedings of the ASME 2014 International Mechanical Engineering Congress and Exposition (IMECE2014); 14-20 November 2014. Montreal. Quebec: IMECE; 2014. DOI: 10.1115/IMECE2014-39077

[38] Dell R et al. 3G security camera network using a thermoelectric generator powered by a geothermal steam pipe in Iceland. Geothermal Resources Council Transactions. 2017; **41**:691-703

[39] Dell R et al. Thermoelectric powered security systems in Iceland using a geothermal steam pipe as a heat source. Proceedings. 2018;**2**:440. DOI: 10.3390/ICEM18-05309

Chapter 6

Heat Recovery and Power Generation Using Thermoelectric Generator

Luis Vitorio Gulineli Fachini, Pedro Leineker Ochoski Machado, Larissa Krambeck, Romeu Miqueias Szmoski and Thiago Antonini Alves

Abstract

In this chapter, experimental analysis of the direct conversion of thermal energy into electric energy was carried out, in order to encourage the conscious use of energy and to reduce waste. The conversion of thermal energy into electrical energy occurs in a thermoelectric generator through the Seebeck effect. This effect is associated with the appearance of an electric potential difference between two different materials, placed in contact at different temperatures. This relation between temperature and electrical properties of the material is known as thermoelectricity. This experimental study has as objective the obtaining of operating characteristic curves of the thermoelectric generator TEG1-12611-6.0, for different temperature gradients and under constant pressure between the heater plate and the heat sink. Resistors were used to heat the thermoelectric generator, which simulates the residual heat, and insulation material to minimize the dissipation of heat to the environment. For cooling, a heat exchanger was used in order to maximize the temperature difference between the sides of the thermoelectric generator. In this way, it was possible to perform an experimental analysis of the obtained electric power for different temperature ranges between the faces of the generator and, with this, verify the applicability in real systems.

Keywords: waste heat recovery, thermoelectric generator, Seebeck effect, thermal energy

1. Introduction

In 2017, total world energy consumption was approximately 13,511 million-ton equivalent of petroleum (MTEP). With the fast industrial growth of developing nations over the last decade, the industrial sector consumed approximately 2852 MTEP. It is estimated that in 2035, the world consumption of energy will increase by more than 30% [1].

Approximately 33% of the total energy consumed in the industry is rejected as residual heat, presenting as a major problem the fact that the most of this rejected energy is identified as low-quality residual heat [2]. This type of waste heat has a

IntechOpen

small working potential, and the temperatures are below 230°C, which implies a low energy density [3]. Concurrently with the concern for global warming and the issues of diminishing oil consumption, there is a strong incentive for the development of more efficient and clean technologies for heat recovery and energy conversion systems using waste heat.

In order to minimize the waste of energy with residual heat, energy recovery systems have been more explored. These systems can become an important object of research and/or application if, at least, part of the thermal energy expelled by industrial equipment to the atmosphere can be reused [4]. In this context, experimental analysis of the direct conversion of thermal energy into electric energy, using thermoelectric generators, was carried out.

The Seebeck effect is related to the appearance of a difference of electric potential between two different materials, placed in contact, however, at different temperatures [5]. Basically, this is the same effect that occurs in thermocouples, where two different materials are connected and submitted to a temperature difference, causing a potential difference to be generated and translated into a temperature reading. In addition to this application, the thermoelectric effect can be explored in the generation of energy for wristwatches and aerospace applications or, even, in the generation of electric energy from the heated gases released in the internal combustion of engines, boiler gases, and/or the geothermal sources. The thermoelectric generators (TEG) have as main characteristics the reduced dimensions, easy adaptation in complex geometry, and very low maintenance [6]. Its conversion efficiency is about 5%; however, studies conducted at the NASA laboratory have reached 20% efficiency for high temperatures [2].

The studied thermoelectric generator consists of an arrangement of small blocks of bismuth telluride (Bi_2Te_3) doped with *n*-type and *p*-type, mounted alternately, electrically in series, and thermally in parallel between two plates of good thermal conduction [7], as shown in **Figure 1**.

The top of the *p-n* junction is heated, and the bottom of the set is cooled; in this way, a temperature gradient is generated. The free electrons of the *n*-type doped elements and the interstices of the *p*-type elements begin to move toward the cold part, that is, the lower part of the system. In the cold part, the *n*-type doped elements acquire negative polarity, while the *p*-type elements get positive polarity. Closing the circuit between the *p-n* elements, an electric potential is generated [7], and, with the electron accumulation at the cold side, an internal electric field is created, causing the Seebeck voltage.

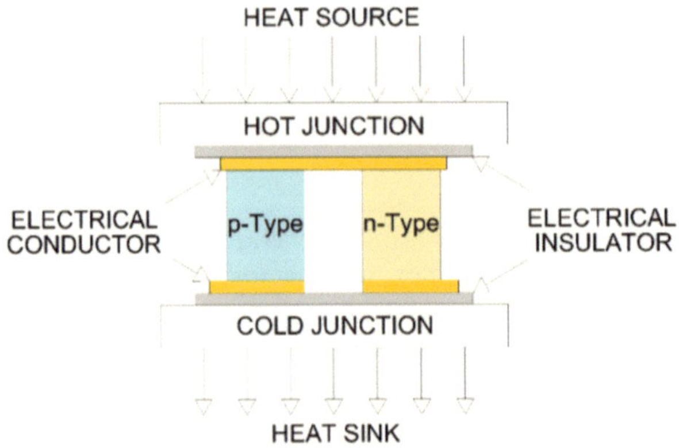

Figure 1.
Schematic diagram of a thermoelectric generator.

2. Methodology

The experimental apparatus and procedure developed for this research are described in details in this section.

2.1 Experimental apparatus

The experimental bench developed to obtain the thermoelectric generator characteristic curve, shown in **Figure 2(a)**, consisted of a laptop (*Dell*™), an uninterruptible power supply (*NHS*™), an aluminum block containing electrical resistors in cartridge, a thermoelectric generator (TEG1-12611-6.0), a water-cooled heat exchanger, a controlled automated resistive load variation system controlled by an *Arduino*™, data logger (*Agilent*™ 34970A with 20 channels), a power supply unit (*Politerm*™ 16E), an ultrathermostatized bath (*SOLAB*™ SL-130), and a variable area flowmeter with throttle (*Omega Engineering*™ FL-2051). In **Figure 2(b)**, the heating and cooling system of the thermoelectric generator and the data acquisition system to obtain generated power by the thermoelectric generator TEG1-12611-6.0 are shown in detail.

The thermoelectric generator used in this experiment is made of bismuth telluride (Bi_2Te_3) and has dimensions of 56 mm by 56 mm with a height of 3.3 mm, totaling a surface area of 0.003136 m^2. An illustration of the generator and, also, its main specifications can be seen in **Figure 3** and **Table 1**, respectively.

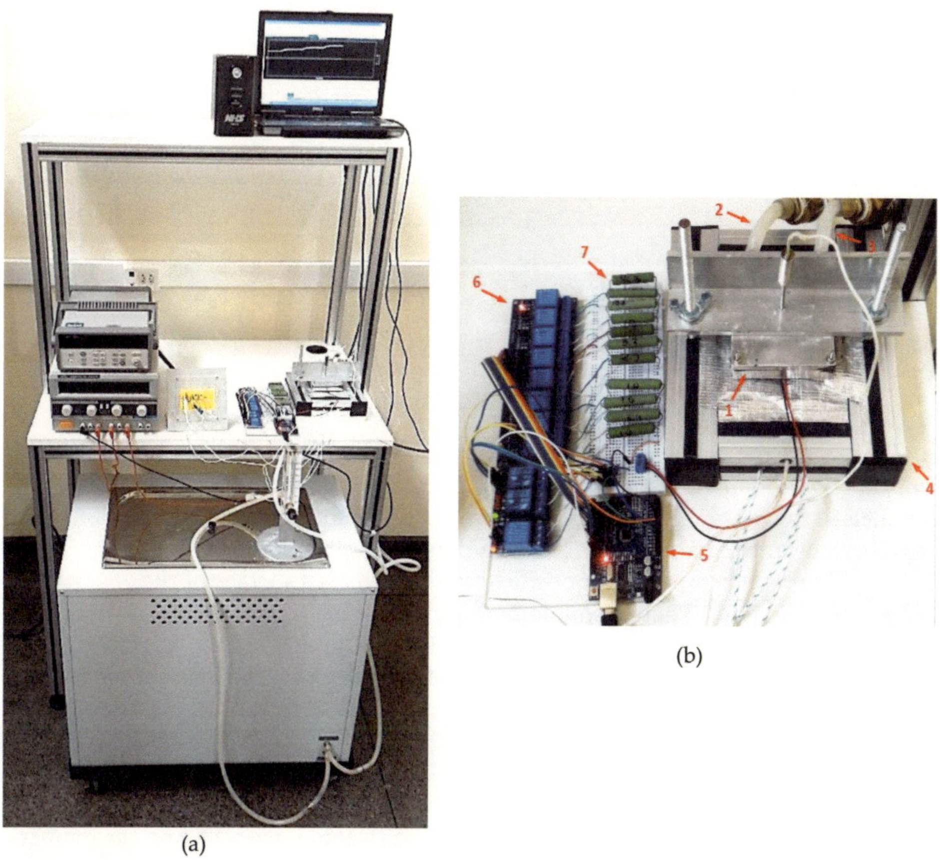

Figure 2.
Experimental apparatus. (a) Experimental bench. (b) Test section and data acquisition system.

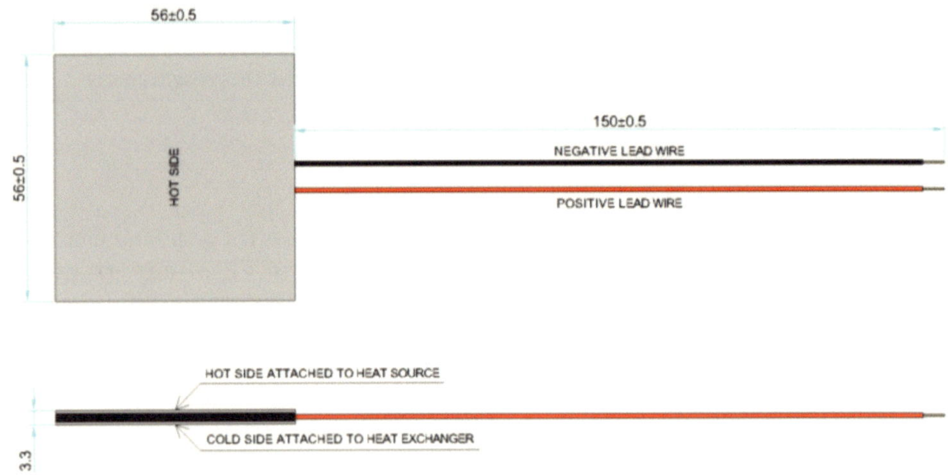

Figure 3.
Illustration of the thermoelectric generator TEG1-12611-6.0.

Parameters	Value	Unit
Hot-side temperature	300	[°C]
Cold-side temperature	30	[°C]
Open-circuit voltage	8.4	[V]
Matched load resistance	1.2	[ohms]
Matched load output voltage	4.2	[A]
Matched load output current	3.4	[A]
Matched load output power	14.6	[W]
Heat flow across the module	Approximately 365	[W]
Heat flow density	11.6	[W/cm^2]
AC resistance under 27°C at 1000 Hz	0.5 to 0.7	[ohms]

Table 1.
Main specifications of the thermoelectric generator TEG1-12611-6.0.

To measure the temperatures of the thermoelectric generator, K-type thermocouples with mineral insulation *Omega Engineering*™, fixed inside machined channels in the heater block and the water-cooled heat exchanger, from the ultrathermostatized bath, were used. The thermocouples were positioned in order to obtain the temperatures of the heated and cooled surface of the generator (Thermocouple 1 and Thermocouple 2—**Figure 4**), resulting in the obtaining of the temperature difference between these surfaces. **Figure 4** shows the schematic diagram of the experimental system used to obtain the characteristic curve.

2.2 Experimental procedure

For the performance of the experimental tests, the ambient temperature was maintained at 16°C ± 1°C by thermal conditioning system *Carrier*™, and the temperature of the water in the ultrathermostatized bath was also maintained at 16°C ± 1°C.

As can be seen in **Figure 4**, the thermoelectric generator was positioned between the heat exchanger and the heating system. The heat exchanger consists of an

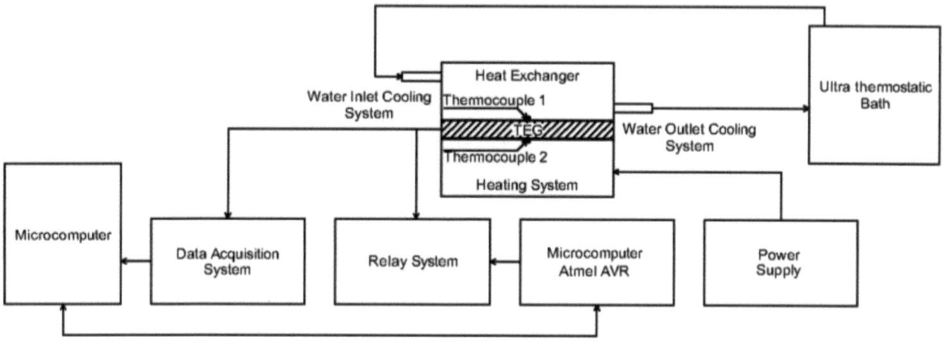

Figure 4.
Schematic diagram of the data acquisition in the experimental apparatus.

aluminum block with machined channels inside. Water is fed by an ultrathermostatized bath with a flow rate of 1 L/min, passes through the channels in order to exchange heat with the upper surface of the TEG, and returns to the ultrathermostatized bath. The heat exchanger, the water inlet, and the water outlet in the exchanger can be observed in items (1), (2), and (3) in **Figure 2(b)**, respectively.

The heating system consists of an aluminum block located inside the base of the experimental apparatus, item (4) in **Figure 2(b)**. The block contains two cartridge resistors associated in parallel, which are responsible for the heating of the block and, consequent, the dissipation of the heat to the TEG. Each cartridge resistor has power of 200 W. In order to increase the heat exchange between the surfaces, the heating system was insulated on its sides using aeronautic thermal insulation. In addition, an aluminum bracket and two threaded rods were used to exert constant pressure in the whole system, increasing the contacts between heat exchanger - TEG and TEG - heating system. These contacts will ensure the required temperature gradient in order to get the Seebeck voltage.

The tests were performed by varying the potential difference (pd) applied to the heating system resistances, which simulate the residual heat, in order to obtain thermal loads from 40 to 320 W, with a variation of 40 W. Each thermal load was maintained for 20 minutes, where the *quasi*-permanent operating condition was reached. To ensure the data reliability, the experimental tests were repeated 3 (three) times for each dissipated power, and errors were compared taking into account that the differences between the mean values which were less than 0.5°C.

The temperature data of the hot and cold surfaces of the TEG were recorded every 10 seconds using the two thermocouples and the data acquisition system, which are saved by the software *Agilent™ Benchlink Data Logger 3*. To obtain information about the thermoelectric generator, a data acquisition system based on *Arduino™*, presented as item (5) in **Figure 2(b)**, was used. The system was responsible for switching resistive loads and obtaining values of voltage and current and, consequently, the power produced by the thermoelectric generator.

For this, an *Arduino™* programming was done to work with a set of relays and a bank of resistors, which supports up to 10 W of power, and the resistances are in a range of 1–50 Ω. In **Figure 2(b)**, the set of relays and the bank of resistors are presented as items (6) and (7), respectively. Different parallel associations between the resistors were activated, obtaining distinct loads that the thermoelectric generator was subjected. As a result, the values of voltage and current are acquired as experimental data and, consequently, the corresponding power. This system was only activated when the condition of the thermal system reached the *quasi*-permanent condition.

Parameters	Measuring instrument	Unit	Uncertainty
Input current	Power source	[A]	± 0.01
Input temperature	Ultrathermostatized bath	[°C]	± 1.00
Other temperatures	K-type thermocouple	[°C]	± 1.27
Input voltage	Power source	[V]	± 0.01
Generated voltage	*Arduino*™	[V]	± 0.01

Table 2.
Uncertainties of direct measurements.

Power of the power supply unit [W]	Uncertainty [W]
40	± 0.38
80	± 0.53
120	± 0.64
160	± 0.72
200	± 0.83
240	± 1.01
280	± 1.29
320	± 1.72

Table 3.
Uncertainties of indirect measurements.

2.3 Experimental uncertainties

The analysis of experimental uncertainties aims to quantify the validity of the data and their accuracy and, therefore, to allow the estimation of the random error present in the experimental results. The error is defined as the difference between the actual value and the indicated value [8].

The experimental uncertainties present in this research were associated to direct and indirect measurements, shown in **Tables 2** and **3**, respectively, and they were calculated according to the error propagation method described in [9]. The direct uncertainties are those relative to the parameters obtained using a measuring tool, being the current, input voltage and temperature, the voltage generated, and other temperatures. The uncertainty of indirect measurement is calculated in the function of other parameters, and a specific tool did not measure it. In the case of this experiment, the only indirect uncertainty is related to the power supplied by the power supply unit to the cartridge resistors, in the heating system. The power quantity was obtained by multiplying the voltage and current supplied by the source.

3. Results and discussions

First of all, the water flow through the heat exchanger, the power dissipated by the electric resistances, and the constant ambient temperature are fixed. After obtaining the *quasi*-permanent condition, the data collection and treatment were performed. **Figure 5** illustrates the output current (I_{out}) versus the output voltage (V_{out}), respectively, in [mA] and [mV], for different dissipated powers in the heating resistors.

It may be noted that the output current and voltage increase according to the temperature difference between the sides of the thermoelectric generator. It is evident from the analysis of **Figure 5** that there is high linearity obtained in the results. It may be further noted that curves have similar slopes; this means that the internal resistance of the thermoelectric generator changes minimally when the operating temperature is varied.

In **Figure 6**, for each temperature difference value, the thermoelectric generator has different internal resistance values. It is possible to note that the resistance values increase in a quadratic form with the rise of the temperature difference.

A curve adjustment with coefficient of determination (R^2) was performed close to 1. The value of the internal resistance of the generator as a function of the temperature difference (ΔT) between the sides of the thermoelectric generator can be defined by

$$\mathrm{Res_{int}} \cong 1.092E - 3\Delta T + 0.515 \quad (1)$$

where Res_{int} is the internal resistance of the thermoelectric generator.

Therefore, the characteristic curve of the thermoelectric generator studied can be expressed by

$$U_g \cong E - (1.092E - 3\Delta T + 0.515)I \quad (2)$$

where E is the electromotive force of the electric generator, U_g is the output voltage of the generator, and I is the electric current.

Figure 7 shows the open-circuit voltage (V_{opn}) and closed circuit voltage (V_{clc}) as a function of temperature differences (ΔT). The open- and closed circuit voltage increase linearly with the rise of the temperature difference. The points obtained for the closed circuit voltage are points where the maximum power generated is obtained.

The curve fit for the open-circuit voltage (U_{opn}) as a function of the temperature difference (ΔT), with a determination coefficient of 0.996, can be determined by

$$U_{opn} \cong 21.664\Delta T - 343.270 \quad (3)$$

where ΔT is the temperature difference between the heated and cooled surfaces of the thermoelectric generator and U_{opn} is the open-circuit voltage of the generator. When the internal resistance of the generator is close to zero, it is assumed that the electromotive force of a generator is equal to the open-circuit voltage. Therefore, by substituting Eq. (3) in Eq. (2), the approximate voltage values are provided as a function of current for this thermoelectric generator, in a range of ΔT between 30°C and 230°C.

$$U_g \cong 21.660\Delta T - (1.092E - 3\Delta T + 0.515)I - 343.260 \quad (4)$$

In this way, it is possible to compare the data obtained experimentally with the theoretical values. **Figure 5** illustrates theoretical curves obtained by Eq. (4). Comparing the experimental and theoretical data, an excellent correlation between the data can be observed, showing that this equation satisfactorily represents the thermoelectric generator behavior, for a temperature difference between 30 and 230°C.

Figure 8 shows the behavior of the output power (P_{out}) as a function of the output voltage (V_{out}) parameterized in the temperature difference between the faces of the thermoelectric generator. It can be observed that with the increase in temperature difference between the faces of the thermoelectric generator, the power generated is also raised.

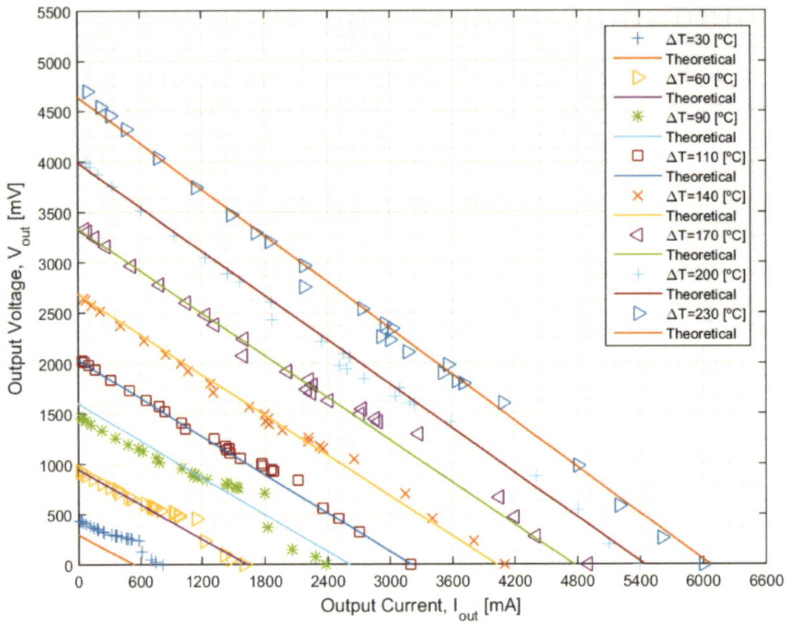

Figure 5.
Output voltage versus output current for different power dissipations.

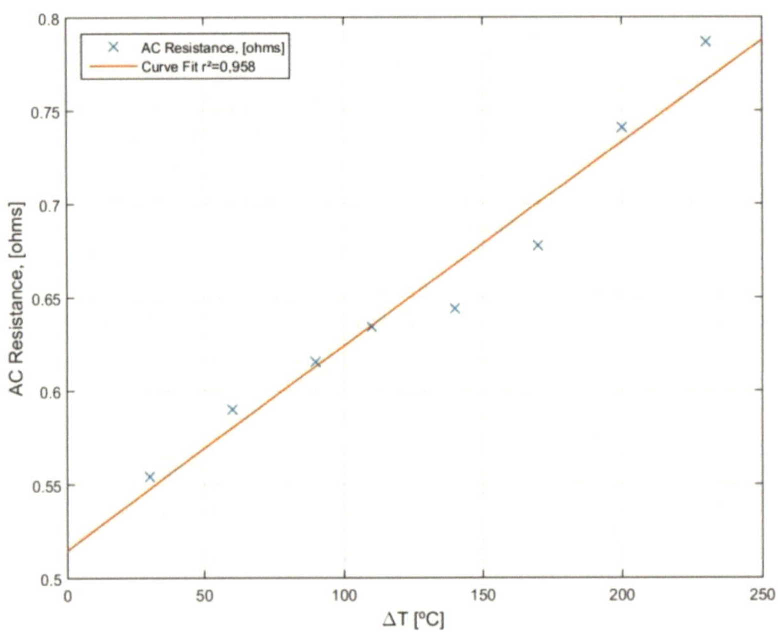

Figure 6.
Internal resistance of the thermoelectric generator.

As expected, the characteristic curves obtained follow a highly quadratic behavior, indicated by the coefficient of determination R^2 close to 1. The maximum power generated occurs when the resistance of the external load is equal to the resistance of the internal load. At this point, the power of 7068 mW is generated with a voltage of 2340 mV for the temperature difference of 230°C.

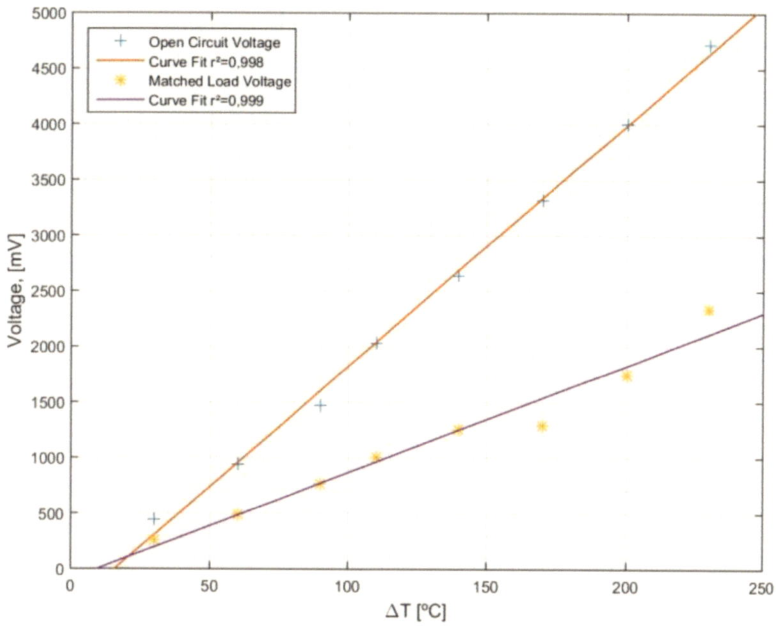

Figure 7.
Open-circuit voltage and closed circuit voltage versus temperature difference.

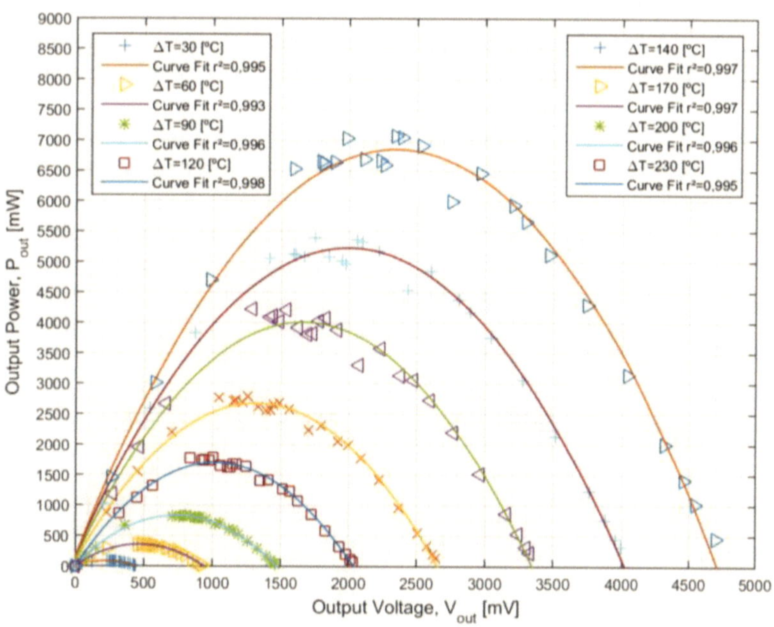

Figure 8.
Output power versus output voltage for different temperature differences.

Figure 9 indicates an alternative way to illustrate the output power data: output power (P_{out}) as a function of output current (I_{out}) parameterized in temperature differences (ΔT) between the surfaces of the thermoelectric generator. It can be seen, again, that the maximum power point is 7068 mW for a current of 3020 mA considering a temperature difference of 230°C.

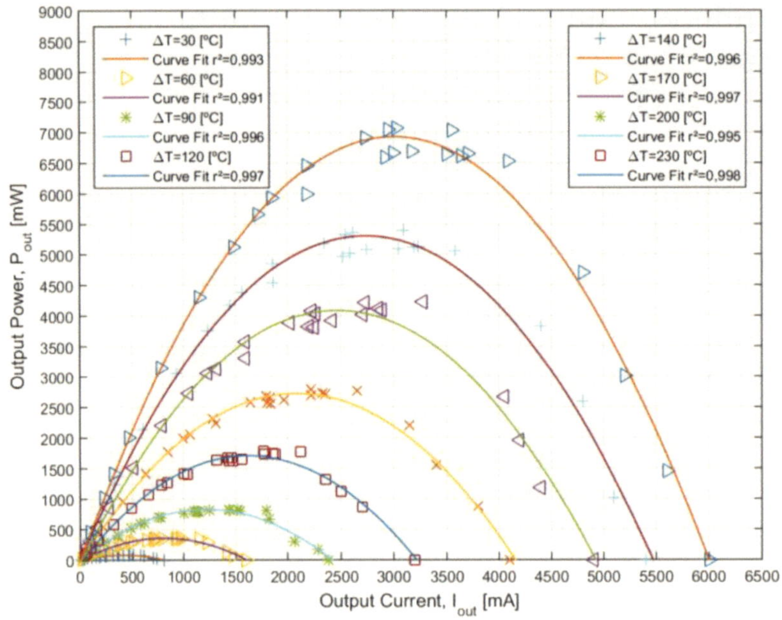

Figure 9.
Output power versus output current for different temperature differences.

Another important point to analyze is the comparison of the amount of power generated by different types of generators. For this, it is of extreme relevance to consider the size of the generators and the generated power density and not just the amount of generated power [W]. As a result, it is important to evaluate how much power is generated per square meter of the generator [W/m^2]. Thus, **Table 4** shows the power generated by the thermoelectric generator area as a function of the temperature gradient.

As shown in **Table 4**, it can be seen that in a condition such as that found in processes with residual heat release, with a temperature gradient in the range of 60°C, the thermoelectric generator is capable of generating 114.14 W/m^2. This value is almost the produced power by a photovoltaic solar panel of monocrystalline cells (*Siemens*™ SM46 [10]), for example, which has A-level efficiency and can generate up to 152.60 W/m^2. In addition, the thermoelectric generator can support higher temperature gradients and, if exposed to more extreme

Temperature difference [°C]	Generated power per area [W/m^2]
30	25.65
60	114.14
90	262.75
110	567.12
140	888.71
170	1345.38
200	1723.54
230	2253.93

Table 4.
Generated power in relation to temperature difference per area.

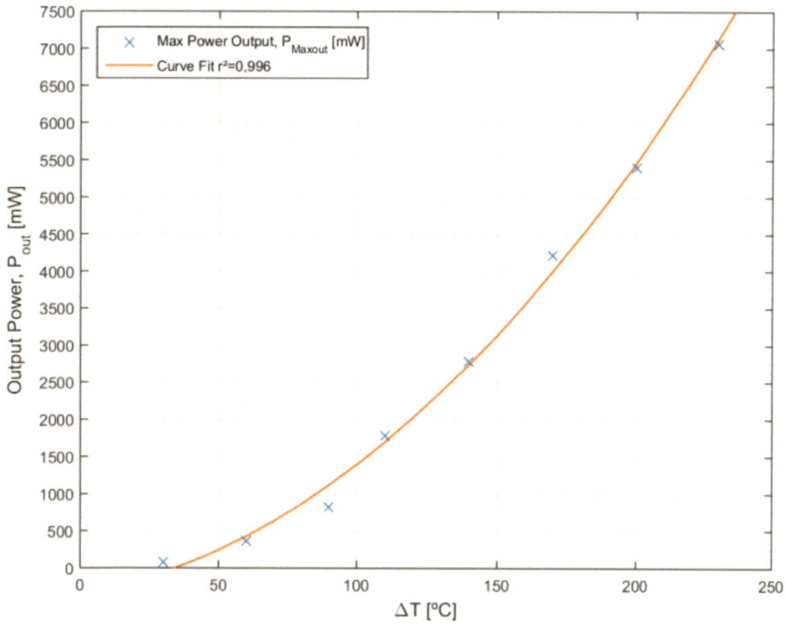

Figure 10.
Output power versus temperature difference.

conditions, can generate up to 2253.93 W/m^2 of electrical power for a temperature gradient of 230°C.

Figure 10 shows the behavior of the output power (P_{out}) as a function of the temperature difference (ΔT). It may be noted that with the increase in temperature difference between the surfaces of the thermoelectric generator, the power generated raises in a quadratic form.

Furthermore, the output power is limited only by the operating temperature of the thermoelectric generator, which is 270°C. Eq. (5) correlates the values of temperature difference for the electric power with a coefficient of determination of 0.996, for temperature differences greater than 7°C.

$$P_{out} = 0.117\Delta T^2 - 5.534\Delta T - 327.308 \quad (5)$$

4. Conclusions

In this chapter, experimental analysis of the direct conversion of thermal energy into electric energy was carried out. An experimental analysis was performed to obtain the operating characteristics of a thermoelectric generator of bismuth telluride (Bi_2Te_3). For this, it was necessary to develop an experimental apparatus to provide the necessary operating conditions. Therefore, it was possible to obtain the operating curves of the thermoelectric generator for a temperature difference between the surfaces of the thermoelectric generator of 30 and 230°C. It is noted that the highest power values delivered by the thermoelectric generator were for the greater temperature differences. Normalizing the power generated by the photovoltaic solar panel and the thermoelectric generator, in conditions close to the real ones of use, both have values of generated power, around 152.60 and 114.14W/m^2, respectively. Also, the thermoelectric generator can reach values of 2253.93 W/m^2 for a temperature gradient of 230°C. In conclusion, the application of thermoelectric generators in the recovery of residual heat is a great instrument to be explored. This kind of device is compact, requires very low maintenance, and has a geometry that can be coupled in most of the industrial systems.

Acknowledgements

Acknowledgments are provided to the CAPES, the CNPq, the PROPPG/UTFPR, the DIRPPG/UTFPR, the PPGEM/UTFPR/Ponta Grossa, and the DAMEC/UTFPR/Ponta Grossa.

Nomenclature

E	electromotive force of the electric generator [mV]
I	electric current [mA]
P	power [mW]
Res	resistance of the thermoelectric generator [ohms]
R^2	coefficient of determination
ΔT	temperature difference [°C]
U	output voltage of the generator [mV]
V	voltage [mV]

Subscripts

out	output
int	internal
g	generator
opn	open circuit
clc	closed circuit

Author details

Luis Vitorio Gulineli Fachini, Pedro Leineker Ochoski Machado, Larissa Krambeck, Romeu Miqueias Szmoski and Thiago Antonini Alves*
Federal University of Technology (UTFPR), Parana, Brazil

*Address all correspondence to: thiagoaalves@utfpr.edu.br

References

[1] BP. Statistical Review of World Energy June 2018. 67th ed. London, UK: BP; 2018

[2] Fleurial JP. Thermoelectric power generation materials: Technology and application opportunities. Journal of the Minerals, Metals and Materials Society. 2009;**61**(4):79-85. DOI: 10.1007/s11837-009-0057-z

[3] UNEP. Thermal Energy Equipment: Waste Heat Recovery. Energy Efficiency. 2006. Available from: www.energyefficiencyasia.org [Accessed: July 1, 2018]

[4] Monreal J. Thermoelectrics: Material advancements and market applications [thesis]. Cambridge, US: Massachusetts Institute of Technology; 2007

[5] Lee H. Thermal Design Heat Sinks, Thermoelectrics, Heat Pipes, Compact Heat Exchangers, and Solar Cells. New Jersey: John Wiley & Sons; 2011. 646 p

[6] Ismail BI, Ahmed WH. Thermoelectric power generation using waste-heat energy as an alternative green technology. Recent Patents On Electrical Engineering. 2009;**2**:27-39

[7] Singh BSB, Saoud A, Remeli MF, Ding LC, Date A, Akbarzadeh A. Design and construction of a simple thermoelectric generator heat exchanger for power generation from salinity gradient solar pond. Jurnal Teknologi. 2015;**76**(5):21-24. DOI: 10.11113/jt.v76.5527

[8] Ismail KAR. Técnicas Experimentais em Fenômenos de Transferência. Mendes Gráfica e Editora: Campinas, BRA; 2000. 488 p

[9] Holman JP. Experimental Methods for Engineers. 8th ed. New York, USA: McGraw-Hill; 2011. 768 p

[10] Siemens Solar Industries. Solar module SM46. Available from: www.siemen.co.uk/sm46.html [Accessed: September 22, 2018]

Section 3

Energy Harvesting Applications of Thermoelectric Devices

Chapter 7

Thermoelectric Control of Deep UV LED to Improve Optical Performance

Pablo Fredes, Ulrich Raff, Ernesto Gramsch, Javiera Cuenca and Jose Pascal

Abstract

A thermoelectric control system using thermoelectric cooler devices (TEC) combined with an aluminum heat dissipater and a fan heat extractor allows improving considerably the optical performance of deep UV LEDs (285 nm) operating at desired temperatures. A proportional, integral, and differential controller (PID) control technique was implemented to control the voltage in the TEC devices, and therefore, the desired range of temperatures can be achieved. The PID parameters are obtained with computational simulations based on physical models and experimental data recordings of the temperature, using a thermistor sensor for the temperature measurements and SiC photodiode with UV enhanced system for the optical power measurements. The experimental data show that decreasing the temperature of the UV-C LED light source using a TEC increases the optical output power, while it has been shown that the lifetime of the LED devices can be improved.

Keywords: thermoelectric cooler, optoelectronics, ultraviolet, light-emitting diode, temperature

1. Introduction

Manufactures of new solid state light sources have focused the attention to deep UV light sources (200–300 nm), where UV-C LEDs (or deep UV-LEDs) have shown great potential in many technological applications, such as water and surface disinfection processes, biomedical instrumentation systems, high density optical recording, lithographic micro fabrication, and biophotochemical research [1–4]. The complexity of a reliable UV-C LED performance depends upon optical efficiency, adequate lifetime, and wavelength accuracy that all concern technological applications. It is known that the UV-C LEDs have lower external quantum efficiency (EQE) with a value less than 10%, compared with conventional LED light sources, for example, blue LED (400–450 nm) has an EQE > 70%. In microbiological research, the shift of the peak wavelengths would result in an inaccurate interpretation in microbial studies [5].

2. Efficiency droop and optical performance of UV-LEDs

The optical performance of the UV-LED devices has been studied by Kheyrandish et al. showing that higher solder point temperatures (T_S) resulted in lower optical output power, shift of the peak wavelength, and an increase of the full width at half maximum (FWHM) [5]. The achieved internal quantum efficiency (IQE) of UV-C LEDs can be quite high, that is, somewhere between 15 and 70%, supported by photoluminescence studies and ray-tracing calculations. However, the external quantum efficiency and the radiant efficiency of UV-C LEDs are below 2% (highest for 280 nm LEDs) [6]. Nevertheless due to increasing development efforts, the UV-C LED devices have much lower output power from single chip than the widely used blue LEDs.

The optical performance of a UV-C LED device decreases, due to a variety of key factors such as quantum-confined Stark effect, carrier delocalization, and others discussed by Versellesi et al. [7]. It is possible to make a classification of droop mechanisms in two levels. The first classification level reflects nonradiative losses taking place inside the optical active region and the second level reflects nonradiative losses taking place outside the optical active region.

Until recently, the principal cause of efficiency droop was unclear in AlGaN LEDs (from blue to UV-C sources). The Auger effect has been proposed as the main cause of efficiency droop in InGaN LEDs. Thanks to Iveland's contribution with direct measurement of Auger electrons emitted from LEDs under electrical injection of high currents, it was possible to conclude that the efficiency droop phenomenon finalized a long-standing controversy [8].

2.1 Thermal stress and lifetime performance

The excessive self-heating of the UV-C LED reduces the lifetime of the devices, the optical performance, and yield spectral shift of the emission. Lifetime limiting issues of UV-C LED occur with high thermal stress in the chip, resulting from the higher electrical current (100–350 mA). The device lifetime may be difficult to assess, in a rigorous fashion. Nevertheless, one recent published result has shown that the lifetime decreases more with a current of 350 mA than with 100 mA in UV-C LEDs by Nikkiso Co, LTD [9]. The thermal extraction is one of the challenges to improve the performance of the LED devices [10]. To resolve this challenge, the thermal conductivity of the package must be considered. The relation between thermal conductivity and thermal resistance is known, and therefore, thermal resistance provided by the manufacturer will be considered.

2.2 LED package and junction temperature

Various indirect techniques and models have been reported for junction temperature T_J measurement techniques including micro Raman spectroscopy, threshold voltage, thermal resistance, photo thermal reflectance microscopy (PRM), electroluminescence, and photoluminescence [11]. In this work, the junction temperature T_J has been obtained applying a thermal resistance technique and direct measurements of the solder point temperatures T_S [12]. The UV-C LED manufacturer allows to compute the optical efficiency of our 285 nm UV-C LED model: VPS171 by Nikkiso Co. The maximal operational solder temperature value can be determined using the maximal junction temperature provided by the manufacturer. The solder point temperature T_S and the junction temperature T_J are linearly related, as shown in (1), where I_{LED} is the electric LED current and V_{LED} is a forward LED voltage. Wall plug efficiency (WPE) will be defined in the next section.

$$T_J(t) = T_S(t) + R_{J-S}(1 - WPE)\, I_{LED}\, V_{LED} \tag{1}$$

3. Wall plug efficiency and temperature dynamics

The wall plug efficiency (WPE) is defined as the ratio of electrical power consumption P_{LED} and optical output power of the UV-C LED device denoted by P_{LIGHT}. Kneissl related the WPE with external quantum efficiency (EQE) and the rate between optical and electrical energy [13] as shown in the next equation.

$$WPE = \frac{P_{LIGHT}}{P_{LED}} = \eta_{EQE}\frac{h\nu}{eV} \tag{2}$$

where h is the Planck's constant, ν is the frequency of the UV light, e is the elementary electric charge, V is the voltage, and η_{EQE} is the external quantum efficiency factor.

The ratio of the number of photons emitted from the LED per second to the number of carriers (electrons) injected into the device per second is precisely EQE. The data sheet of our 285 nm UV C LED showed that at 350 mA and 6 V, the UV-C LED can provide an optical output power of 30 mW. This implies that the WPE is 1.43% and EQE is 1.97%, that is, that 1.97% of the injected electrons is converted to UV-C light. Eq. (2) shows that the optical output power P_{LIGHT} is related with P_{LED} and the temperature, and its relation with EQE is shown in Eq. (3)

$$P_{LIGHT}(T_J) = \eta_{EQE}(T_J)\frac{h\nu I_{LED}}{e} \tag{3}$$

Eq. (3) allows to relate the EQE with the junction temperature. Since the junction temperature can be controlled, the EQE factor (η_{EQE}) can be improved.

Joule's first law shows the relation between the heat generated by an electric current passing through a conductor. Eq. (4) shows that the electrical power is converted to light and heat.

$$P_{LED} = I_{LED}\, V_{LED} = P_{LIGHT} + P_{HEAT} \tag{4}$$

The light power P_{LIHT} can be determined using the electrical efficiency factor WPE, obtained with the parameters of the UV-C LED provided by the manufacturer as a linear relation, $P_{LIGHT} = WPE\ P_{LED}$.

As shown in **Figure 1**, P_{HEAT} is divided into two heat transfer rates. *P1* is the heat transfer rate between the junction surface and the ambient environment, and *P2* is the heat transfer rate between the junction surface and the cooling system. For operations at ambient temperature, P_1 is negligible compared to P_2.

3.1 Newton's law of cooling

The thermal performance of the UV-LED device shows that the solder temperature dynamics $T_S\ (t)$ can be modeled with Newton's law of cooling, where T_{EST} represents the temperature in the steady state.

$$\frac{dT_S}{dt} = -K(T_S - T_{EST}) \tag{5}$$

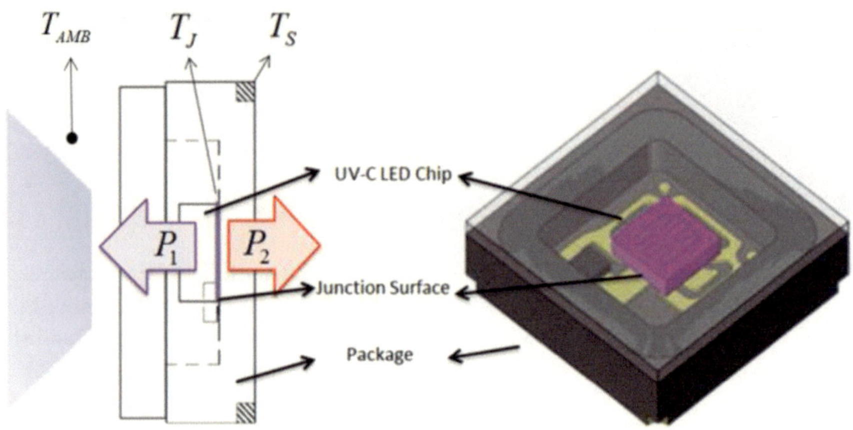

Figure 1.
UV-C LED chip inside the package represented in 2D and 3D diagrams. The size of the LED package shown at the right-hand side is 3.6 × 3.6 mm. T_{AMB} is the ambient temperature, T_J is the junction temperature, and T_S is a solder point temperature.

Taking into account that the cooling process begins at ambient temperature T_{AMB}, commonly in the range of 19–25°C, to finally achieve the steady state of the temperature T_{EST}. The analytical solution for the solder temperature is

$$T_S(t) = T_{EST} + (T_{AMB} - T_{EST})\, e^{-Kt} \quad (6)$$

Commonly, the inverse of the constant K is referred to as time constant of the system. The determination of the constant K of Newton's law of cooling is important and has to be determined experimentally for each LED device (including thermoelectric cooler system). The K value depends on the predefined stationary temperature value T_{EST}. The value of the cooling temperature constant K will be able to model the thermal performance of the system. Therefore, the UV-C LED device temperature can be controlled applying a corresponding strategy.

4. Experimental setup

Our setup is composed by three essential parts: the UV-C LED device, the optical sensor (SiC photodiode model: SG01D-C18, from SGLux), and the temperature sensor (calibrated thermistor). The UV-C LED system is mounted on an optical table (**Figure 2B**). The light output is directed toward an optical sensor at a predefined distance, and finally, a temperature sensor is mounted at the solder point. Experimental data are acquired using a computational interface.

The thermoelectric cooler cell (TEC) is composed of a large number of PN and NP semiconductor arrays. TEC modules can convert voltages (V_{TEC}) into a temperature gradient, based on a phenomena discovered by Peltier in 1834. In this study, a TEC model number TEC1-12706 by Hebei I.T. Co. Ltd. was used.

The experimental setup for the measurements of solder temperature dynamics T_S consists of a 10 kΩ thermistor model NTC Murata, which is directly put in contact on the solder point. The temperatures are measured at the edge of the LED package shown in the right-hand side of **Figure 1**. The thermistor is powered by a stabilized voltage source device, model: LM2596 DC-DC. The signal of the thermistor is connected to an A/D converter to acquire and record (every second) the temperature data.

Figure 2.
Components of the UV-C LED device (A): (a) UV-C LED SMD chip, (b) MCPCB, (c) aluminum plate, (d) thermal isolators, (e) thermoelectric cooler device (TEC), (f) aluminum heat dissipater, and (g) fan heat extractor. (B) All components shown in (A) are mounted.

The accuracuty of our thermistor is 1%, specified by the manufacturer, our operation range is from −40 to 60°C, therefore the maximum absolute error is ε1 = 1°C. The computational interface error is ε2 = 0.5°C. It is possible to estimate the standard error of the mean (SEM) for the temperature measurements, using the standard deviation of 20 repetitions of the data recorded for a fixed value of temperature, monitored by a thermal camera model FLIR E 63900 with an emissivity setting of 0.95. The standard deviation was S = 0.73°C, and the total error for the temperature recording can be estimated, that is, ε3 = 0.16°C. Finally, to complete the total error of the solder point temperature measurements T_S, we sum all previously mentioned errors Δ = ε1 + ε2 + ε3 = 1.66°C. Previous work shows that the standard deviation between the modeled and true junction temperatures for blue LEDs was only 3.5°K [14].

5. Thermal performance and empirical modeling

5.1 Thermal resistance

The thermal partial resistances correspond to the various functional groups in the system or heat path and characterize their individual thermal behavior (**Figure 3**). The total thermal resistance R of the system is then equal to the sum of the individual thermal resistance components.

$$R = R_{J\text{-}S} + R_{S\text{-}B} + R_{B\text{-}TEC} + R_{TEC\text{-}AL} + R_{AL\text{-}AMB} \tag{7}$$

$R_{J\text{-}S}$ is the thermal resistance of the UV-C LED chip and describes the transfer of heat within the LED housing (junction-to-solder-point), $R_{S\text{-}B}$ is the thermal resistance of the solder point to the metal core printed circuit board (MCPCB), $R_{B\text{-}TEC}$ is the thermal resistance of the MCPC and aluminum plate to the thermoelectric cooler device (TEC), $R_{TEC\text{-}Al}$ is the thermal resistance of the TEC to the aluminum heat dissipater, and finally $R_{AL\text{-}AMB}$ is the thermal resistance of the aluminum heat dissipater to the ambient environment. The thermal resistance between board and TEC ($R_{B\text{-}TEC}$) is not strictly a thermal resistance, but it is technically a heat engine, which can be considered a peculiar thermal resistance with negative value [15]. It is clear that this is an advantage of the TEC, since it can be observed in (7), that

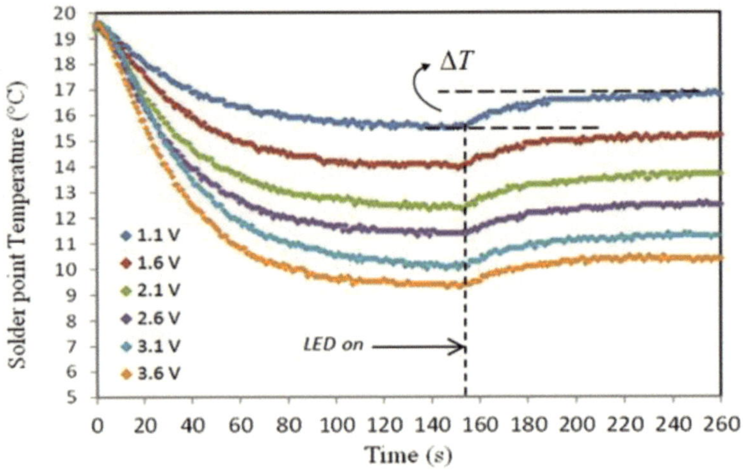

Figure 3.
Performance of solder temperature as a function of time for different values of TEC voltages. The dotted vertical line shows the time when the UV-C LED is turned on after the first stationary temperature obtained from the TEC. The difference between first and final stationary temperatures is given by ΔT.

the negative value of $R_{B\text{-}TEC}$ reduces the total thermal resistance of the device. As a general rule, the smaller the thermal resistance of a LED, the faster the heat can be dissipated from the housing.

5.2 Thermal performance

To obtain the stationary temperature of the system T_{EST} and its relationship with TEC voltage, we apply a pre-cooling turning on of the LED. First, the TEC is turned on with a predefined constant voltage value. Once the stationary temperature is reached, the UV-C LED source is turned on. The system will then reach its final stationary temperature.

T_{EST} is related with V_{LED} and T_{AMB}. It is possible to achieve the stationary temperatures in the system making the cooling process with the TEC before turning on the LED. It is known that the relation between TEC voltage and gradient temperature is linear. The linear relation is described in the following equation.

$$T_{EST}(V_{TEC}, T_{AMB}) = T_{AMB} - mV_{TEC} \tag{8}$$

If T_{AMB} is constant, it is possible to obtain the value of slope m, relating the empirical recording of the cooler process dynamics, for different values of V_{TEC}. The value of m is a thermal property of each system, in a specific range of temperatures. For example, the data displayed in **Figure 3**, allow to confirm the linear relationship stipulated in Eq. (8), that is, $T_{EST} = 19.5–1.59V_{TEC}$. In the six curves shown in **Figure 3**, V_{LED} was maintained constant.

6. Temperature control using TEC and PID controller

The control technique implemented in this application has been developed applying the previously described physical model, the results of the simulations, and the recorded experimental temperature data.

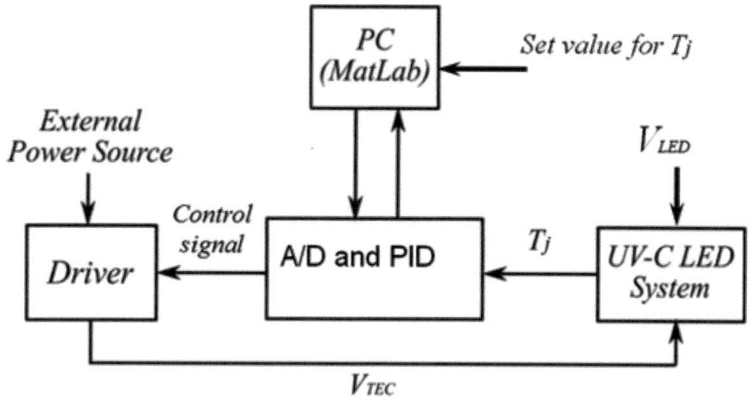

Figure 4.
Schematic diagram for the temperature controller.

According to Eq. (1), the solder point temperature can be obtained knowing the junction temperature. We can define a reference junction temperature $T_{J\text{-}REF}$ and obtain a stationary reference solder point temperature $T_{EST\text{-}REF}$. The controller operates comparing $T_{EST\text{-}REF}$ to $T_S\ (t)$ to obtain an error signal that is fed to the PID controller. The PID controller applies V_{TEC} to the system to yield $T_{EST\text{-}REF}$.

The block diagram displayed in **Figure 4** is implemented with the MATLAB Simulink software, considering the empirical values of the time constant, the ambient temperature T_{AMB}, the increment ΔT, the linear relation between V_{TEC} and T_{EST} shown in Eq. (8), and the set values for the junction temperatures $T_{J\text{-}REF}$.

7. Results

A computational tuning based on a simulated plant is used to obtain the PID parameters, which were then applied and verified experimentally. For each steady-state temperature T_{EST}, the PID parameters can be obtained in the simulations to be consequently applied in the controller (**Figure 5**).

Using these gain parameters for the PID controller applied to our system, we can fit the simulations of the temperature dynamics for three different steady-state operations. When the UV-C LED is turned on, the control of the system increases the TEC voltages to recover the starting temperature states as displayed in **Figure 6**.

One of the major advantages of the temperature control is the improvement of the optical output power. This result is displayed in **Figure 7**. The relation between junction temperature and UV intensity shows that as the temperature drops, the UV intensity increases. The plotted data demonstrate that reduction of the junction temperature T_J from 59 to 44°C increases the UV irradiation by 6%.

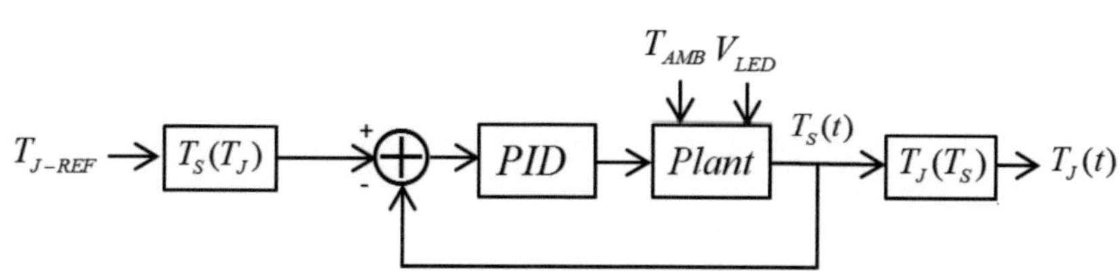

Figure 5.
Block diagram for PID controller of T_J based on T_S control.

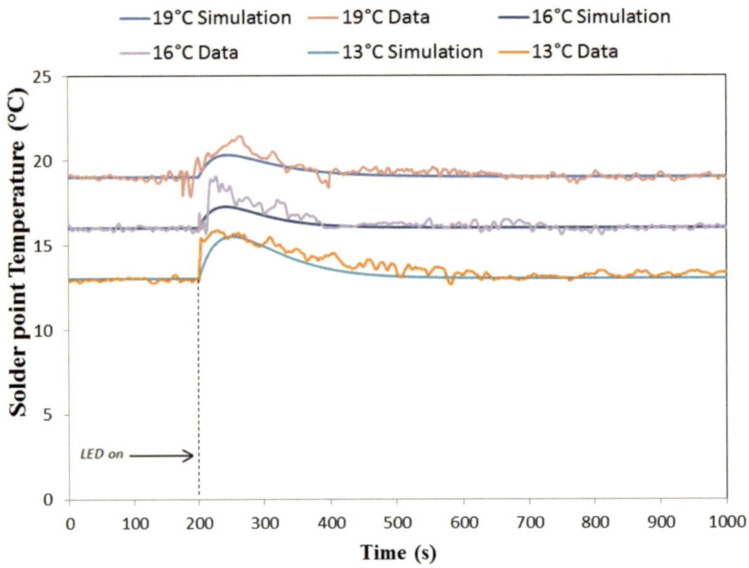

Figure 6.
Fit of the simulations and experimental temperature dynamics for three different steady-state operations after initial precooling.

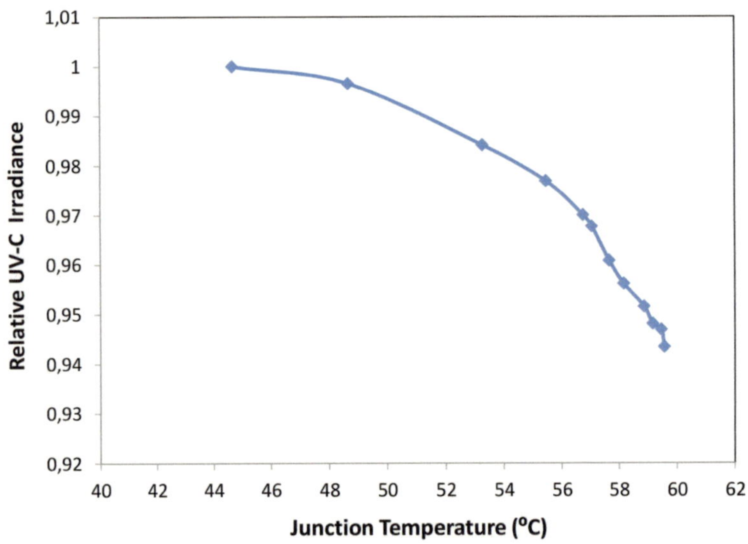

Figure 7.
Normalized irradiance as a function of junction temperature.

8. Conclusions

The result shows that the TEC can effectively reduce the thermal resistance of the high power UV-C LED device and therefore substantially increase the optical performance. In addition, the lifetime might be highly improved [8]. Since the system temperature affects the optical performance, it is essential to control it. The TEC device makes it possible to implement an appropriate control strategy. Each LED device has different thermal performance, depending on many factors, for example, thermal conductivity of the heat dissipaters, the fan extractor velocity, and operation ambient temperature. At lower temperature operations, optical

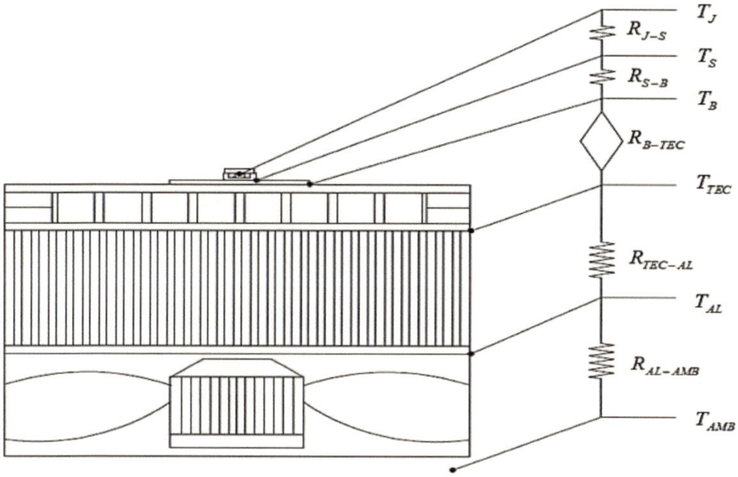

Figure 8.
Representation of a UV-C LED system, describing the series of thermal resistances and the points of the thermal interfaces, specified on the right hand side.

performance is improved. The operation of the UV-C LED at lower temperatures implies that the intrinsic lifetime could be substantially improved. The implementation of the TEC device for the thermal control of the UV-C LED light source allows an efficient optimization of the optical performance (**Figure 8**).

Acknowledgements

Portions of this work were presented at the OSA Light, Energy and Environment Congress in November 2017, titled "Improving the Optical Performance and Life Time of Deep UV LEDs by Thermoelectric Control". This work was supported in part by for the Hydraluvx Spa, Santiago, Chile and Mrs. Andrea Bravo, Hydraluvx's Graphics Designer, Santiago Chile is gratefully acknowledged for the rendering of **Figures 2** and **8**.

Author details

Pablo Fredes*, Ulrich Raff, Ernesto Gramsch, Javiera Cuenca and Jose Pascal
Department of Physics, Universidad de Santiago, Chile

*Address all correspondence to: pablo.fredes@usach.cl

References

[1] Khan A, Balakrishnan K, Katona T. Ultraviolet light-emitting diodes based on group three nitrides. Nature Photonics. 2008;**2**:77-84

[2] Sholtes KA, Lowe K, Walters GW, Sobsey MD, Linden KG, Casanova LM. Comparison of ultraviolet light-emitting diodes and low-pressure mercury-arc lamps for disinfection of water. Journal of Enviromental Technology. 2016;**37**:2183-2188

[3] Hirayama H, Maeda N, Fujikawa S, Toyoda S, Kamata N. Recent progress and future prospects of AlGaN-based high-efficiency deep-ultraviolet light-emitting diodes. Japanese Journal of Applied Physics. 2014;**53**:100209

[4] Inoue S, Tamari N, Taniguchi M. 150 mW deep-ultraviolet light-emitting diodes with large-area AlN nanophotonic light-extraction structure emitting at 265 nm. Applied Physics Letters. 2017;**110**:141106

[5] Kheyrandish A, Mohseni M, Taghipour F. Development of a method for the characterization and operation of UV-LED for water treatment. Water Research. 2017;**122**:570-579

[6] Schur MS, Gaska R. Deep-ultraviolet light-emitting diodes. IEEE Transactions on Electron Devices. 2010;**57**:12-25

[7] Verszellesi G, Saguatti D, Meneghini M, Bertazzi F, Goano M, Meneghesso G, et al. Efficiency droop in InGaN/GaN blue light-emitting diodes: Physical mechanisms and remedies. Journal of Applied Physics. 2013;**114**:071101

[8] Pernot C. Improving Deep UV LED Performance. Presented at the LED China Conference, Shanghai, China; 2016. http://www.semi.org.cn/semiconchina/mail/160331/1/02-SEMICONCHINA-DUV-LED-cyrilpernot.pdf

[9] Nikkiso Co. Standard Specifications for UV LED, model VPS131, Vol. 1.2; 2016

[10] Kneissl M et al. Development of UV-C LED Disifection. TECHNEAU; 2010

[11] Akca H, Yasa Y, Ayaz R, Durusu A, Ajder A, Nakir I, Tanrioven M. Thermal management of power LED system. In: Proceedings of ICRERA; Milwaukee, WI, USA; 2014. pp. 760-764

[12] Xi Y, Gessmann T, Shah JM, Kim JK, Schuberta EF. Junction and carrier temperature measurements in deep-ultraviolet light-emitting diodes using three different methods. Applied Physics Letters. 2005;**86**:031907

[13] Kneissl M, Rass J. A brief review of III-nitride UV emitter technologies and their applications. In: III-Nitride Ultraviolet Emitters. 1st ed. Vol. 227, Springer Series in Materials Science. 2016. pp. 1-25

[14] Reed ML, Wraback M, Lunev A, Bilenko Y, Hu X, Sattu A, et al. Device self-heating effects in deep UV LEDs studied by systematic variation in pulsed current injection. Physica Status Solidi. 2008;**5**(6):2053-2055

[15] Riffat SB, Ma X. Thermoelectrics: A review of present and potential applications. Applied Thermal Engineering. 2003;**23**(8):913-935

Chapter 8

Harnessing the Automotive Waste Heat with Thermoelectric Modules Using Maximum Power Point Tracking Method

Dario Gandini, Marcello Chiaberge and Andrea Nepote

Abstract

The present work shows a comprehensive methodology and design steps to recover energy from the automotive waste heat. A thermoelectric generator must be connected to a power converter in order to extract the maximum power from the generator and, also, satisfy different constrains to charge a battery. Starting from the electrical model of thermoelectric cells, it is evaluated their combination to realize a thermoelectric generator (TEG) comply with the automotive regulation, then considering input/output electric characteristics, it is evaluated the best converter topology to satisfy all constrains. Design steps and power dissipation estimation are deeply explained. TEG and power converter models are simulated in a model-based environment to allow the design of the control algorithms. The control system consists of nested control loops. Two maximum power point tracking (MPPT) algorithms are evaluated. The MPPT output is used as reference for a current control loop. The maximum power characteristic of a TEG has a quadratic behavior and working without the tracking of the maximum power point could drastically decrease the generated power from the TEG and the system efficiency. There are presented simulation results of the control algorithms and experimental data are shown in order to validate the design steps.

Keywords: automotive, thermoelectric generator, waste heat, storage, DC-DC converter, MPPT

1. Introduction

Nowadays the reuse of wasted energy is an important task for many fields, particularly for the automotive one. This aims at more efficient systems, reducing pollution and making the product more compatible with the environment, regulations, and consumers. Internal combustion engine vehicles do not totally convert the chemical energy into mechanical, the greater part of a typical fuel path is wasted in the coolant system and in the exhaust, and according to Chuang and Chau [1], only 25% of the energy flows into specific vehicle operations (**Figure 1**). The strategy is to convert the heat wasted energy into electricity by means of a thermoelectric generator (TEG), which is based on the Seebeck effect. Heat flows through the TEG that, thanks to the difference of temperature (ΔT) between the hot and cold sides, generates electrical power that flows into the connected load. **Figure 2** shows the quantities that interact with the TEG: heat, temperature, thermal resistance, interface material, and electrical quantities.

The increase of delta temperature also increases the available power, but for each delta temperature, there is only one maximum power that can be generated connecting a precise load. A simple approach is to connect a DC-DC converter after the TEG. The DC-DC converter can set the current flow; thus, from the TEG point of view, it acts as a variable load. Specific control algorithms can be implemented

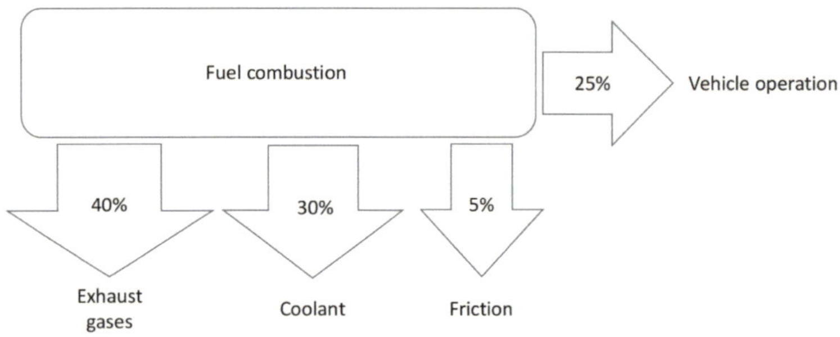

Figure 1.
Quantity of energy used for the vehicle operation.

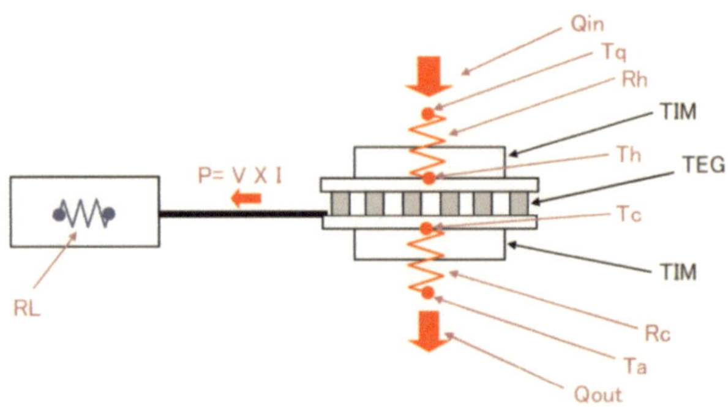

Figure 2.
TEG quantities. Q_{in} is the heat quantity to flow into TEG, Q_{out} is the heat quantity to flow out of TEG, T_q is the temperature of heat source, T_a is the temperature of ambient (cold source), T_h is the temperature of TEG hot-side surface, T_c is the temperature of TEG cold-side surface, TIM is the thermal interface material, R_h is the heat resistance of TIM between TEG and heat source, R_c is the heat resistance of TIM between TEG and ambient, P is the electrical power, V is the voltage, I is the current, and RL is the load resistance.

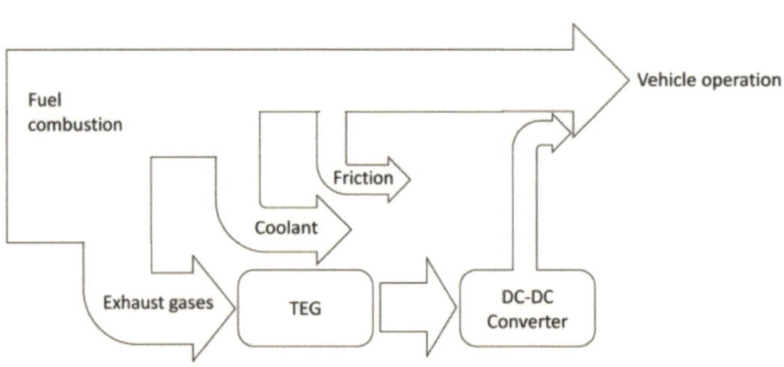

Figure 3.
Diagram of the wasted heat energy recovery system.

into the converter in order for each delta temperature to be the correct load. The converter output can be connected to the vehicle battery, so the generated power can be directly used for vehicle operations or stored for future needs. The described system allows to recover part of the wasted energy increasing the total efficiency of the vehicle. **Figure 3** shows a conceptual diagram of the system from fuel combustion to wasted heat and the recovery system, TEG plus DC-DC, that generates power useful for vehicle operations.

2. Design methodology of an energy recovery system

This chapter describes a comprehensive methodology for the design of an energy recovery system. Starting from physical characteristics of the generator, its electrical model is obtained. Since the generator under examination is a thermoelectric generator, all derived parameters will be expressed as a function of the delta temperature, the temperature difference between the hot and cold sides of the generator. Firstly, it is characterized and modeled a single thermoelectric cell, then it is designed the cells combination to make the TEG generator: the resulting equivalent electrical model of the generator is similar to a single cell with grater voltage and current parameters. The parameters obtained by the generator model are used as constraint for the DC-DC converter topology definition, and the design steps are described, highlighting the necessary trade-offs. After these stages it is possible to face the control algorithms. In order to design a control algorithm, there are two necessary information: what the target is and what the weapons are to reach the target. Finally, the single sections are put together in a single simulation environment where it is possible to test the global system changing temperature and load conditions.

2.1 TEG and its electrical model

Figure 4 shows the open-circuit voltage and the maximum power characteristics of the hypothetical thermoelectric cell. They are represented as a function of the delta temperature across the cell. The maximum allowed delta temperature is 180° C. The open-circuit voltage (V_{OC}) grows linearly with the delta temperature (Eq. (1)), while the available maximum power (P_{MAX}) has a parabolic behavior (Eq. (2)). The maximum open-circuit output voltage is almost 8 V, and the

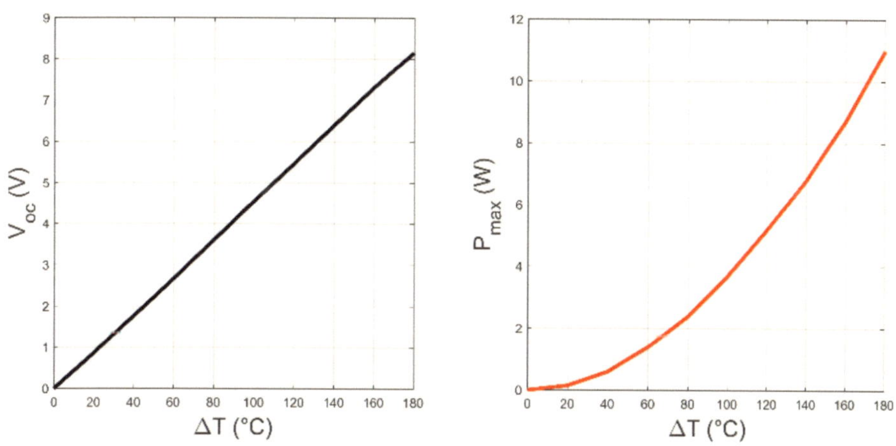

Figure 4.
TEG cell characteristics as a function of delta temperature; on the left the open-circuit voltage and on the right the available maximum power.

maximum power is a little more than 10 W. The coefficients m_v and m_p are specifics of the selected TEG cell:

$$V_{OC} = m_v \cdot \Delta T \quad (1)$$

$$P_{MAX} = m_p \cdot \Delta T^2 \quad (2)$$

Figure 5 shows the voltage-current and the power-voltage characteristics at a single delta temperature. When the TEG output is closed in short circuit, the voltage is zero, and the output current reach is local maximum (I_{sc}). It is important, with the purpose of generating the available maximum power, to know, for each delta temperature, the correct voltage-current couple. In Eq. (3) the relation between the voltage to generate the maximum power (V_{mpp}) and the open-circuit voltage is pointed out, namely, the delta temperature. In Eq. (4) the current to generate the maximum power (I_{mpp}) is derived as a function of the already calculated parameters:

$$V_{mpp} = \frac{V_{OC}}{2} = \frac{m_v}{2} \cdot \Delta T \quad (3)$$

$$I_{mpp} = \frac{P_{MAX}}{V_{mpp}} = \frac{2 \cdot m_p}{m_v} \cdot \Delta T \quad (4)$$

From the voltage-current characteristic, it is also possible to extract the value of the TEG equivalent series resistance (Eq. (5)). **Figure 6** shows the voltage-current characteristic at different delta temperature: different slopes mean different equivalent resistance:

$$R_{eq} = \left| \frac{I_{sc}}{-V_{oc}} \right| \quad (5)$$

Figure 7 shows the equivalent series resistance calculated at different delta temperature. Eq. (6) shows the equivalent series resistance as a function of delta temperature. The equivalent electric model of a TEG cell is a variable voltage

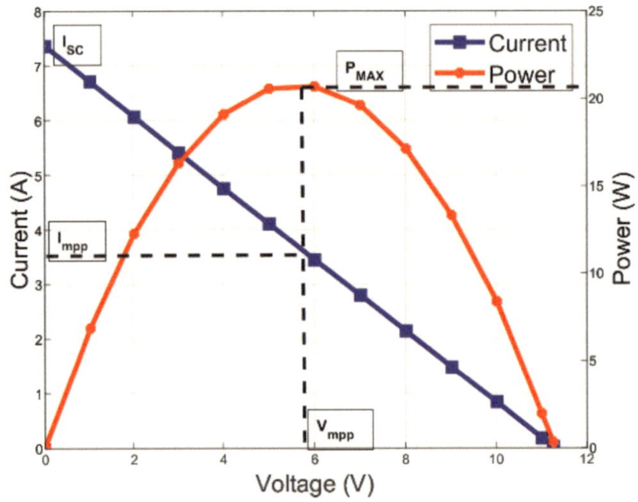

Figure 5.
Current-voltage and power-voltage characteristics.

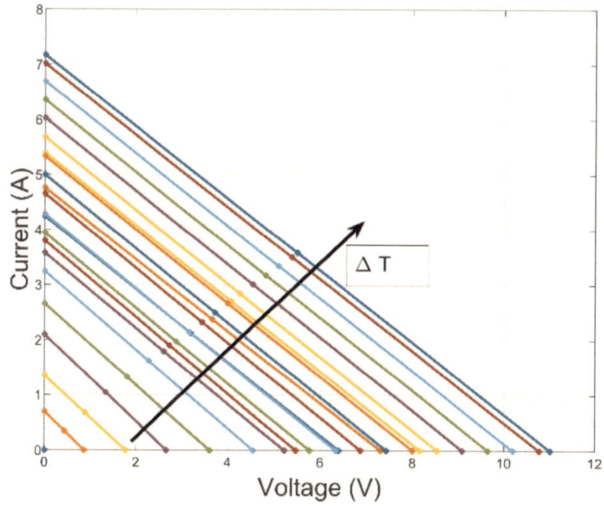

Figure 6.
Equivalent resistance as a function of temperature difference.

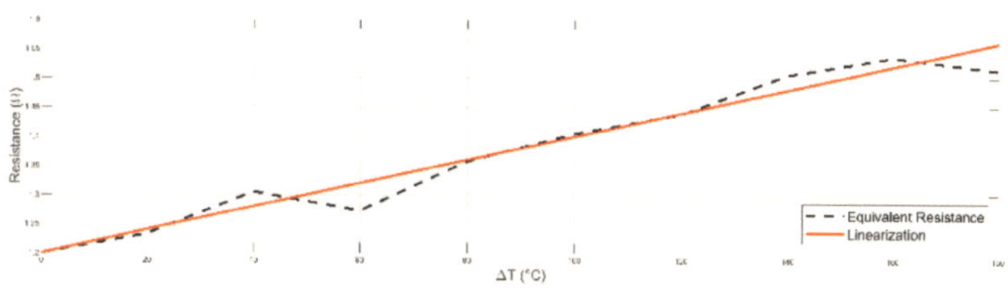

Figure 7.
TEG equivalent resistance as a function of delta temperature.

generator with a variable series resistance, both proportional of the delta temperature:

$$R_{eq} = m_r \cdot \Delta T + q_r \quad (6)$$

More TEG cells can be combined in order to increase the total generated power. Combination can be done by putting the cells in series, increasing output voltage, or, in parallel, increasing the output current. Due to regulatory constraint, it is better to keep the maximum voltage under 60 V. Furthermore, the TEG pack design is strictly related to available space, point of contact of the exhaust pipe, and heat cooling passing through the TEG (it will cause different available powers between near cells) [2, 3]. The TEG pack for this application is constituted by six cells in series and four in parallel. **Figure 8** shows the equivalent electric model of the TEG pack constituted by 24 cells. Under the hypothesis of a good pack design, Eqs. (7)–(12) give a general relation between electrical parameters and the number of cells in series (C_s) and in parallel (C_p):

$$V'_{oc} = V_{OC} \cdot C_s \quad (7)$$

$$R'_{eq} = R_{eq} \cdot \frac{C_s}{C_p} \quad (8)$$

$$P'_{max} = \left(V'_{oc}\right)^2 / R'_{eq} \quad (9)$$

$$I'_{sc} = I_{sc} \cdot C_p \quad (10)$$

$$V'_{mpp} = V_{mpp} \cdot C_s \quad (11)$$

$$I'_{mpp} = I_{mpp} \cdot C_p \quad (12)$$

2.2 DC-DC topology evaluation

Usually the aim of a DC-DC converter is to regulate an unregulated input voltage for a specific load; in this application the final goal is to take the maximum power from the generator at the larger possible temperature excursion, namely, manage a wide input voltage and set the current flows. A battery is connected to the output, so the output voltage must be slightly higher than the battery voltage, in order to allow the recharge. The battery voltage is set at 12 V. The input voltage is supposed to be in the range from 0 to 60 V.

Every DC-DC topology has its own relation between input and output voltages. In this test case, commonly used topologies, like buck or boost converter, are excluded. A buck converter (**Figure 9**) works only with input voltage higher than output; it means only with high delta temperature. A boost converter (**Figure 10**) works only with input voltage lower than output; it means only with low delta temperature. In order to manage input voltage lower and higher than the output, it

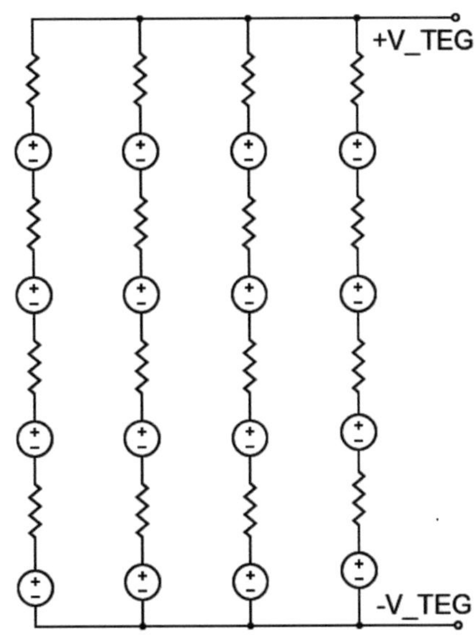

Figure 8.
TEG cell configuration electrical equivalent.

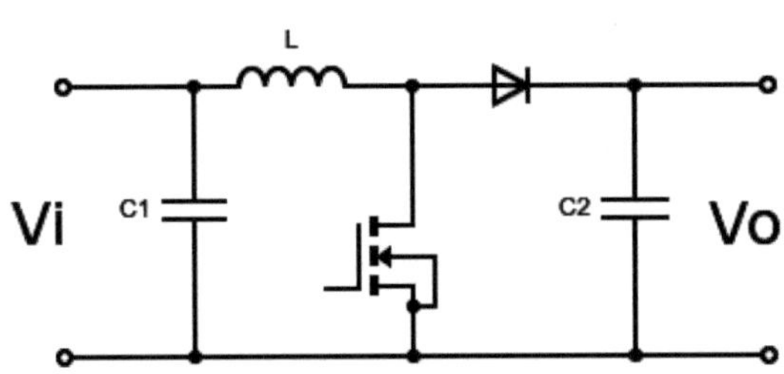

Figure 9.
Boost converter topology.

is possible to combine these two topologies. From the combination of a buck and boost converters, there are two topologies:

- Boost and buck cascaded
- Non-inverting buck-boost (NIBB)

In the boost and buck cascaded topology, there are two stages: the first one is a boost stage, and the second is a buck stage. Both stages have their own inductor (**Figure 11**). If the input voltage is lower than the output voltage, only the first stage works, while the high-side switch belonging to the second stage is always closed; on the other hand, if the input voltage is higher than the output voltage, only the second stage works, while the low-side switch belonging to the first stage is always open. As a third working option, if the input voltage is near the output voltage, both stages could work: the boost stage slightly increases the input voltage, and the buck stage regulates and complies with the desired output voltage.

In the non-inverting buck-boost topology, the two stages are compressed, and there is only one inductor that is shared by them (**Figure 12**). The first leg is

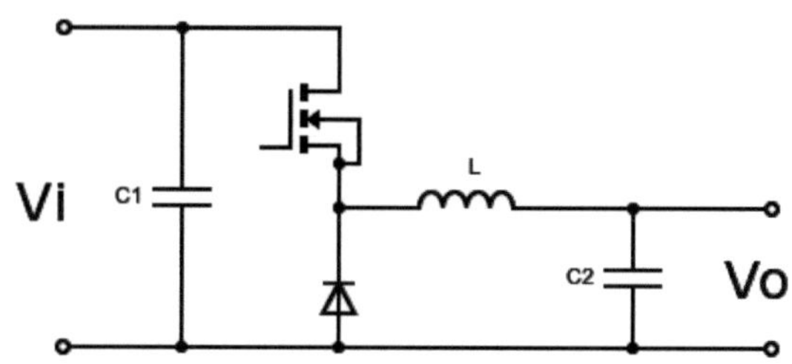

Figure 10.
Buck converter topology.

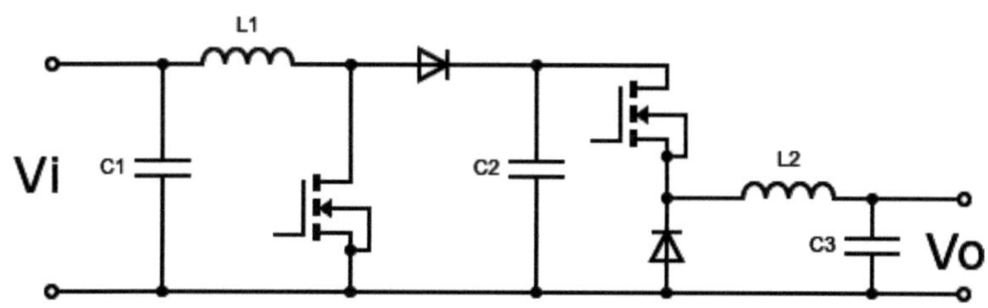

Figure 11.
Boost and buck cascaded topology.

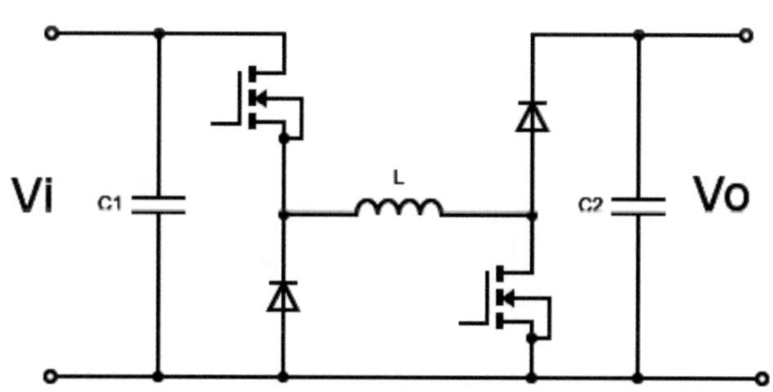

Figure 12.
Non-inverting buck-boost (NIBB) topology.

dedicated for buck operation, while the second leg for the boost operation. As in the previous topology, it is possible to manage input voltage that is lower or higher than the output voltage. The three working regions are:

- Buck (Vi > Vo)
- Boost (Vi < Vo)
- Buck-boost (Vi ≈ Vo)

Practically the inductor is shared only in the buck-boost working region; on the other only one leg is active, and the other has a fixed state. In order to reduce weight, number of components, and loss sources, the NIBB topology is the perfect candidate for this application. In the next section, design steps and loss estimation of the NIBB topology are presented.

2.3 NIBB design

In the non-inverting buck-boost topology, a buck and a boost are fused together and share a single inductor. A technique for improving the converter's efficiency is the synchronous rectification. The two versions of the NIBB converter are also called two-switch and four-switch non-inverting buck-boost. The two-switch NIBB (2SW-NIBB) has diodes, while in the four-switch NIBB (4SW-NIBB), the diodes are replaced by power MOSFETs (**Figure 13**). This topology will be used for the DC-DC converter, and in the following, the design steps are reported.

Below the starting specifications for the design are reported:

- Input voltage $0 < Vi < 60$ V.
- $V_o \approx 12$ V.
- Maximum output power 300 W.
- Output power is proportional to input voltage.

The steady-state relation between input and output voltages of a 4SW-NIBB is outlined in Eq. (13). D_A and D_B are the commands of the switches, namely, duty cycles; their range is from 0 to 1. D_A is the duty cycle of the buck leg high-side switch; the command of the low-side switch is $1 - D_A$ ($\overline{D_A}$). D_B is the duty cycle of

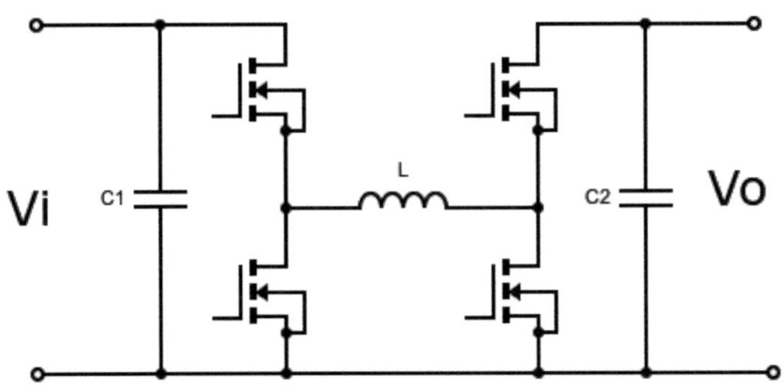

Figure 13.
NIBB with synchronous rectification (4SW-NIBB) circuit.

the boost leg (low-side switch); the command of the high-side switch is $1 - D_B$ ($\overline{D_B}$):

$$V_o = \frac{V_i \cdot D_A}{1 - D_B} \tag{13}$$

It is known from Section 2.1 that to extract the maximum power from the TEG, its output voltage must be half of the open-circuit voltage; in the following V_{mpp} is considered as the effective input voltage of the DC-DC converter. **Figure 14** shows the relation between V_{mpp} and delta temperature, highlighting the working regions. The general relation between input and output voltages could be simplified considering in which region the converter is working (Eq. (14)):

$$\begin{cases} V_o = \dfrac{V_i}{1 - D_B}, D_A = 1 \ (Boost\ region) \\ V_o = \dfrac{V_i \cdot D_A}{1 - D_B} \ (Buck - Boost\ region) \\ V_o = V_i \cdot D_A, D_B = 0 \ (Buck\ region) \end{cases} \tag{14}$$

In order to define switching frequency and inductor value, the two operation modes of a DC-DC converter are introduced: continuous current mode (CCM) and discontinuous current mode (DCM). The DCM operation is characterized by a current of the inductor that is zero during the switching period, while in the CCM, the inductor current never reduces to zero (**Figure 15**). The DCM operation is particularly a disadvantage for the buck region. In order to avoid the DCM operation for the buck working region, Eq. (15) must be verified. The value of L_{crit} is calculated in Eq. (16), where ΔI_L is the current ripple on the inductor and can be considered a percentage of the output current, typically 20–30% of the output current:

$$L > L_{crit} \tag{15}$$

$$L_{crit} = D_A \cdot \frac{V_i - V_o}{\Delta I_L \cdot f_{SW}} \tag{16}$$

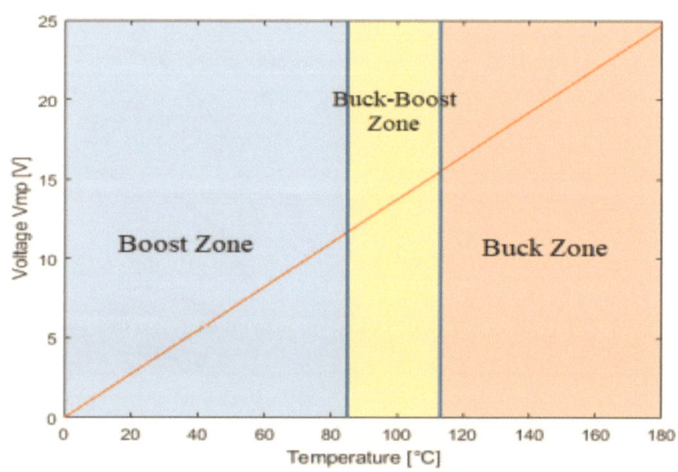

Figure 14.
Converter input voltage as a function of the temperature difference, highlighting working regions.

Figure 16 shows the relation between L_{crit} and the temperature difference considering different switching frequency. The use of high switching frequency allows the use of smaller inductor but means higher switching losses. The switching frequency and the inductor value are a trade-off between reduced component dimension and losses. The last important parameter for the inductor selection is the saturation current that is calculated in Eq. (17), where μ is the converter efficiency that as a safety factor can be set to 80%:

$$I_{L,sat} = \frac{V_o \cdot I_o}{\mu \cdot V_{in}} + \frac{\Delta I_L}{2} \tag{17}$$

The capacitor selection is based on maximum RMS current that is flown into it and the voltage ripple. The output capacitor selection is based on the maximum RMS current in the capacitor during the boost working region and the output voltage ripple specification. Maximum RMS current in the output capacitor is calculated in Eq. (18). Input voltage is the maximum input voltage for the boost working region. The output voltage ripple has two contributions: the first one due to capacitor equivalent series resistance (ESR) and the other related to capacitance (Eq. (19)):

$$I_{Co,RMS} = I_o \sqrt{\frac{V_o}{V_{IN}}} \tag{18}$$

$$\Delta V_{Co} = \frac{I_o V_o}{V_{in}} \cdot ESR + \frac{I_o}{C_o \cdot f_{SW}} \left(\frac{V_o - V_{in}}{V_o} \right) \tag{19}$$

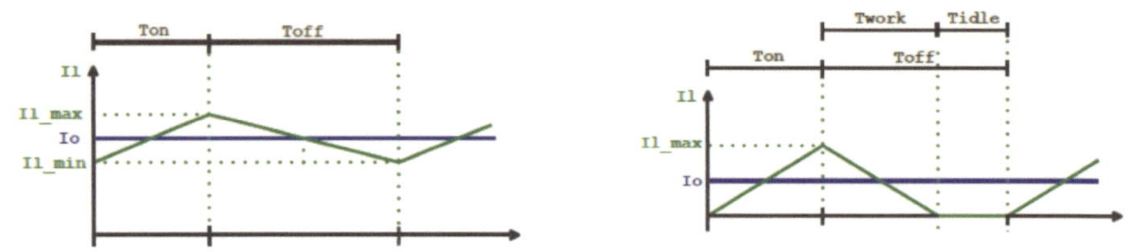

Figure 15.
Inductor current waveform: (left) CCM and (right) DCM.

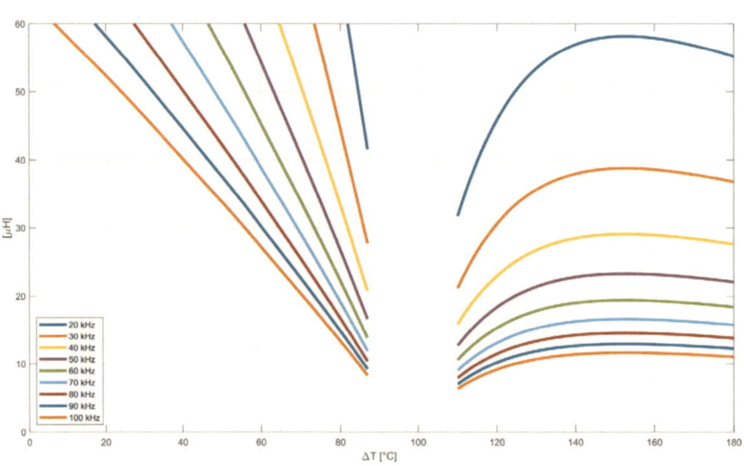

Figure 16.
Minimum CCM inductance as a function of temperature difference.

Input capacitor selection closely follows the method used for the output capacitor. The maximum RMS current flowing in CIN occurs in buck mode that is calculated in Eq. (20). As for the output capacitance, the input voltage ripple has two contributions: the first one related to input capacitor ESR and the second due to capacitance (Eq. (21)):

$$I_{C_{in}, RMS} = \sqrt{\frac{V_o}{V_{IN}}\left[I_o^2\left(1-\frac{V_o}{V_{in}}\right)+\frac{1}{12}\Delta I_L^2\right]} \quad (20)$$

$$\Delta V_{C_{in}} = I_o \cdot ESR + \frac{I_o \cdot V_o}{V_{in} \cdot C_{in} \cdot f_{SW}}\left(1-\frac{V_o}{V_{in}}\right) \quad (21)$$

2.3.1 Loss estimation

At first approximation inductor and capacitors are due to the component's equivalent series resistance (ESR). In Eqs. (22) and (23), the input capacitor losses and output capacitor losses are calculated, respectively; inductor losses are calculated in Eq. (24):

$$P_{C_{in}}(\Delta T) = ESR_{C_{in}} \cdot I^2_{C_{in}, RMS}(\Delta T) \quad (22)$$

$$P_{C_o}(\Delta T) = ESR_{C_o} \cdot I^2_{C_o, RMS}(\Delta T) \quad (23)$$

$$P_L(\Delta T) = ESR_L \cdot I_o^2(\Delta T) \quad (24)$$

The MOSFET losses have two components: conduction and switching losses. Conduction losses are calculated in Eq. (25). R_{dson} is the equivalent resistance of the MOSFET when it is closed. For the switching losses, it is necessary to separate the two legs: Eq. (26) for the buck leg and Eq. (27) for the boost leg. t_R is the rise time and t_F is the fall time, parameters of the selected switch. While output current is a constraint of the application, the switching frequency is a degree of freedom that can be used in order to reduce switching losses (**Figures 17** and **18**):

$$P_{cond}(\Delta T) = 2 \cdot R_{dson} \cdot I_o^2(\Delta T) \quad (25)$$

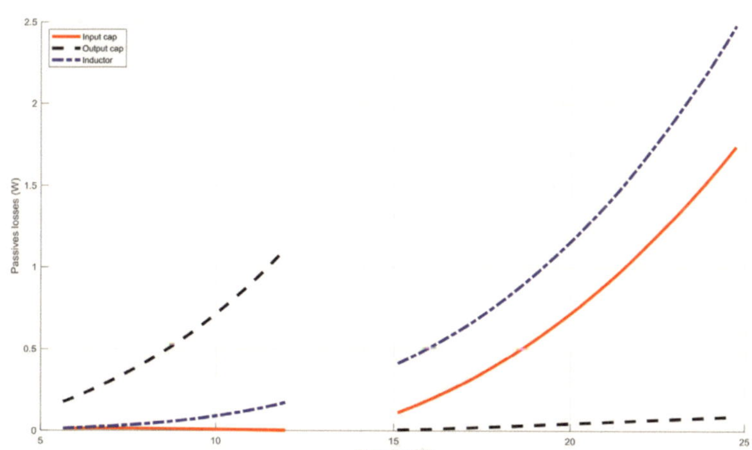

Figure 17.
Loss estimation of the passive components. Continuous line input capacitor; dashed line is the output capacitor; dash-dotted line is the inductor. Parameters of ***Table 1****.*

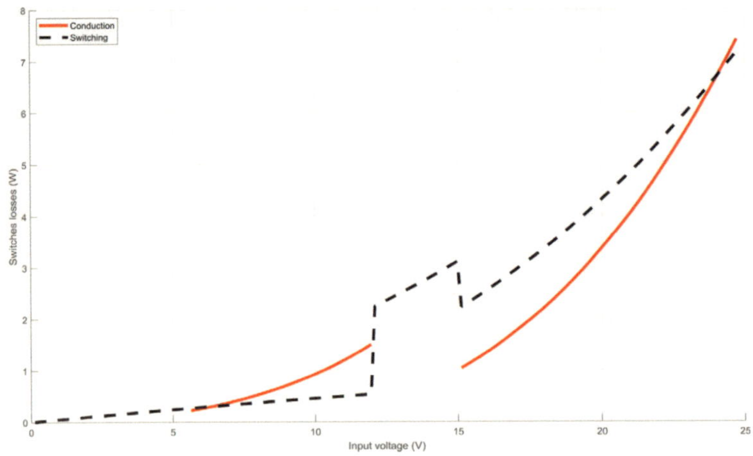

Figure 18.
*Loss estimation of the switches. Continuous line is the conduction losses; dashed lines are the switching losses. Parameters of **Table 1**.*

Parameter/component	Value
f_{sw}	30 kHz
L	30 μH
L ESR	20 mΩ
C_{in}	660 μF
ESR C_{in}	20 mΩ
C_{out}	330 μF
ESR C_{out}	40 mΩ
$R_{ds,ON}$	5 mΩ
t_R	350 ns
t_F	200 ns

Table 1.
Summary of the converter design.

$$P_{sw}(\Delta T) = f_{sw} \cdot (t_R + t_F) \cdot V_{mpp}(\Delta T) \cdot I_o(\Delta T) \tag{26}$$

$$P_{sw}(\Delta T) = f_{sw} \cdot (t_R + t_F) \cdot V_o \cdot I_o(\Delta T) \tag{27}$$

The final design is a trade-off between inductor and capacitor values, namely, dimension and weight, and the losses. **Table 1** shows the summary of the design. Device's parameters are comparable with real components available on the market. These parameters are used for the control algorithms and simulation sections.

2.4 Control algorithms

The control system consists of nested control loops. In order to extract the maximum power from the TEG, two different maximum power point tracking (MPPT) algorithms are evaluated. The MPPT output is used as reference for voltage and current control loops. Analog variables are input and output voltage and current of the converter; commands are the duty cycles of the legs. Input voltage and current are used for the maximum power point tracking; output voltage and current are used for the other control loops.

2.4.1 Maximum power point tracking

There are many MPPT algorithms that are suitable for use with a TEG, most of which have been adapted from those used by photovoltaic systems [4]. The most common are perturb and observe (P&O) and incremental conductance (IC) algorithms [5–8]. P&O algorithm is an iterative algorithm, its working principle is to introduce a perturbation in the operating voltage and current, and then the operating power is observed and compared with the power of the previous step. If the power difference is positive (negative), subsequently the MPP will be reached if the perturbation is stepped in the same (opposite) direction. The flowchart for P&O is shown in **Figure 19**. A P&O algorithm cannot determine if the MPP is reached: when it does reach it, it will pass the operating voltage beyond which will result in a decrease of power; in response the algorithm will reverse the tracking direction. Therefore, in employing a P&O algorithm, the operating point oscillates around the MPP. In order to reduce the oscillation, it is possible to make the perturbation size smaller. However, reducing perturbation size means reducing the algorithm dynamics and increasing the time to reach the MPP. A trade-off between steady-state performance and dynamic response is the use of variable perturbation size [9, 10].

In order to improve the steady-state error, the incremental conductance (IC) algorithm was implemented. It tracks by stepping the voltage like the P&O, and it operates by incrementally comparing the ratio of the derivative of conductance (dI/dV) with the instantaneous conductance (I/V). The advantage of using IC method with respect to the P&O is that it can determine the MPP; in fact, at MPP, the derivative of power with respect to voltage (dP/dV) is 0. The relation between conductance and power-voltage ratio is expressed in Eq. (28). The basic rules for IC are written in Eq. (29). The IC algorithm flowchart is reported in **Figure 20**:

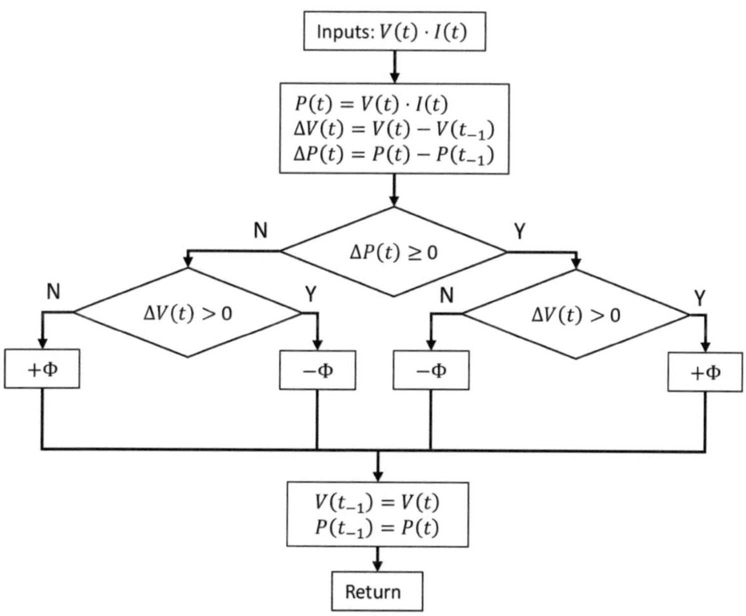

Figure 19.
P&O algorithm flowchart.

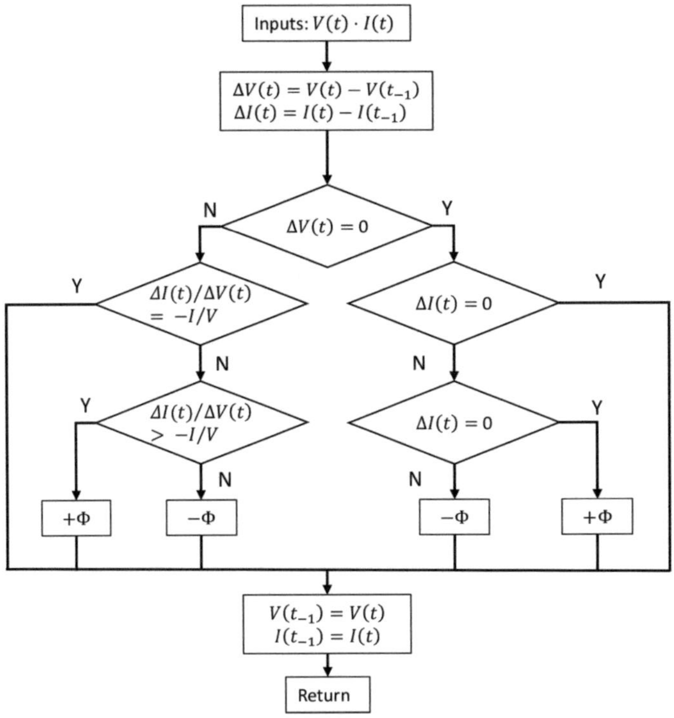

Figure 20.
IC algorithm flowchart.

$$\frac{dP}{dV} = \frac{d(VI)}{dV} = I + V \cdot \frac{dI}{dV} \tag{28}$$

$$\begin{cases} \dfrac{dI}{dV} = -\dfrac{I}{V}, \text{ at MPP} \\ \dfrac{dI}{dV} > -\dfrac{I}{V}, \text{ left of MPP} \\ \dfrac{dI}{dV} < -\dfrac{I}{V}, \text{ right of MPP} \end{cases} \tag{29}$$

In practical application the equality condition rarely exists. Moreover, the conductance instantaneous value is related to perturbation size; because of that a marginal error (ε) is introduced. The IC algorithm considers the MPP reached when the operating point is within a certain error margin (Eq. (30)):

$$\left| I + V \cdot \frac{dI}{dV} \right| < \varepsilon \tag{30}$$

Using nested control system, the output perturbation of the MPPT algorithm is used as a reference for the internal control loops that are described in the next subsection.

2.4.2 Current loop

A change of the TEG operating point and a change of converter output current are alike. In order to control the output current, an average current control loop is used. The output of the MPPT algorithm is used as a reference for the current control loop. The average current of the inductor is related to the output current, and the stationary relation between input and output current is reported in Eq. (31):

$$I_i = \frac{I_o \cdot D_A}{1 - D_B} \tag{31}$$

The system-state variables of a NIBB converter are reported in Eq. (32); the model of the system is expressed in matrixial form in Eq. (33):

$$x = \begin{bmatrix} i_L \\ v_C \end{bmatrix} \tag{32}$$

$$\dot{x} = \begin{bmatrix} 0 & -\overline{D_B}/L \\ \frac{\overline{D_B}}{C} & -1/R \cdot C \end{bmatrix} \cdot x + \begin{bmatrix} D_A/L \\ 0 \end{bmatrix} \cdot V_i \tag{33}$$

The Laplace form of linearized model of the converter is reported in Eq. (34):

$$\begin{cases} i_L(s) = \dfrac{V_{in}}{sL} \cdot d_A(s) + \dfrac{V_o}{sL} \cdot d_B(s) - \dfrac{\overline{D_B}}{sL} \cdot v_o(s) \\ v_o(s) = -\dfrac{I_L}{sC} \cdot d_B(s) + \dfrac{\overline{D_B}}{sC} \cdot i_L(s) - \dfrac{1}{sRC} \cdot v_o(s) \end{cases} \tag{34}$$

The system is characterized by two control inputs (d_A, d_B), but as shown in Section 2.3, it is possible to consider them one at a time if the system is in buck or boost region. In order to have only one control variable also in the buck-boost region, it is possible to fix one leg, i.e., boost leg, and command only the other. Another way is to use the same command: the two duty cycles ($d_A = d_B$). The buck transfer function between duty cycle and current is reported in Eq. (35), and the boost transfer function between duty cycle and current is reported in Eq. (36). The effect of the output capacitor series resistance is neglected:

$$G_{di} = \frac{R}{(1 + sRC_{out})} \tag{35}$$

$$G_{di} = \frac{V_{in}}{(1 - D_B)} \cdot \frac{sRC + 2}{R(1 - D_B)^2 + sL + s^2 RLC_o} \tag{36}$$

The proposed controller for the average current control loop is a simple proportional integrative (PI) controller, the current reference comes from the MPPT algorithm, and the output current is the feedback. **Figure 21** shows the system transfer function and the controller behavior.

2.5 Simulations and design validation

The complete system is simulated in a model-based environment; thermoelectric generator is modeled using the mathematical model described in Section 2.1, while for the DC-DC converter and the load, it is used as a package to use circuital and power components inside the environment; values of **Table 1** are used. Sense blocks are placed before and after the converter, and measures are used in the control block in order to generate the correct commands. The complete system functional blocks are reported in **Figure 22**. For the purpose of validating the single control algorithm, the blocks are tested separately. **Figure 23** shows a detail of the P&O MPPT algorithm simulation: the algorithm's output is the equivalent resistance

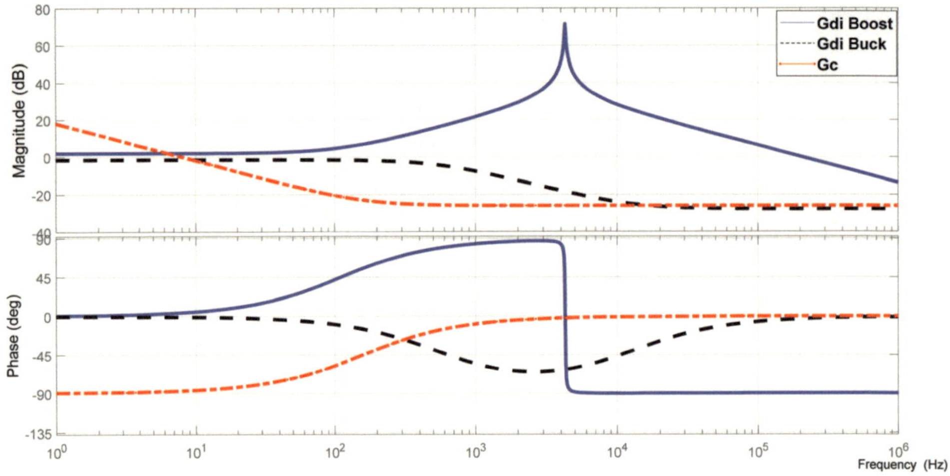

Figure 21.
System transfer functions and controller transfer function.

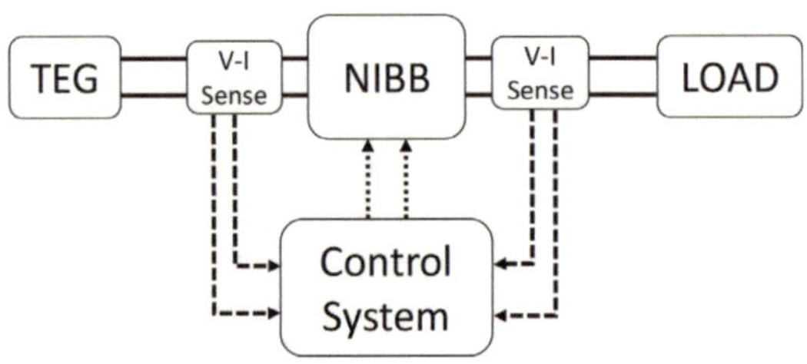

Figure 22.
Wasted heat energy recovery system block diagram.

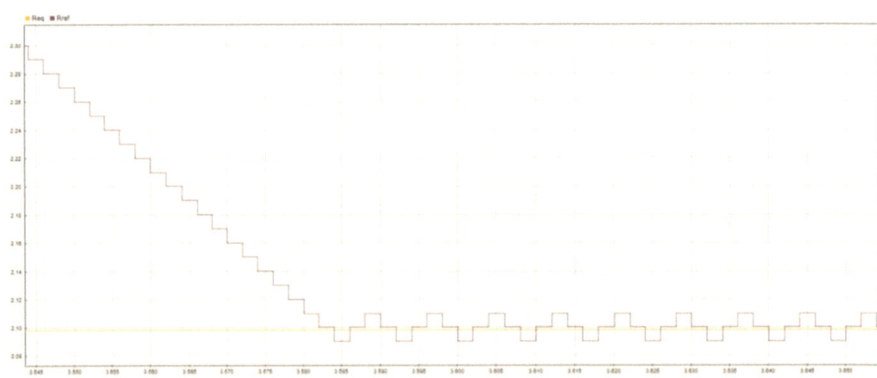

Figure 23.
Maximum point tracking.

placed after the TEG. The operating point is fixed; the value of the resistance seen by the TEG is initialized at a wrong value, after some iteration; and according to the maximum power transfer theorem, the maximum power point is reached. During the steady-state operation, the resistance value continuously changes its value as seen in Section 2.4.1.

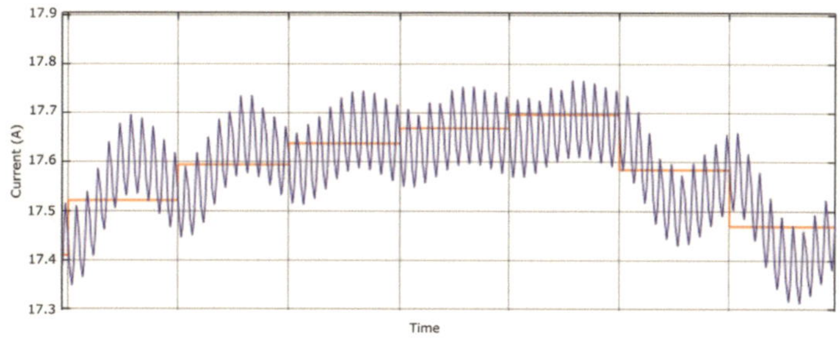

Figure 24.
Average current control.

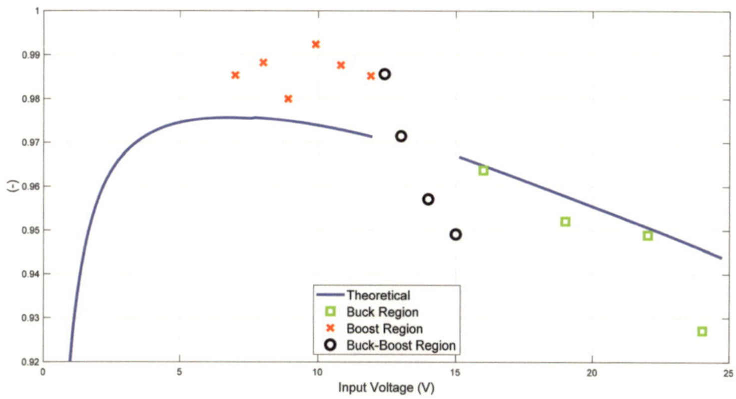

Figure 25.
Efficiency as a function of input voltage.

Figure 24 shows the average current loop behavior with variable reference. The reference changes with steps like the behavior of the MPPT algorithm output. The reported current is the inductor current. In order to realize a nested control loop, it is important that the inner current loop band is larger than the outer MPPT loop.

A NIBB prototype is realized using the designed components. In order to validate the functionality and loss estimation, some power tests are carried out. **Figure 25** shows the overlapping of the theoretical efficiency and the real efficiency obtained with open-loop commands. The difference from real data and estimation is due to the discrepancy of parasitic values: estimation plot is referred to **Table 1**; real selected components have slowly better parameters.

3. Conclusions

A comprehensive methodology for the design of a wasted heat energy recovery system is given. The general approach is to model separately the single stages starting from physical characteristic and derivate the resulting design constraints. Starting from the characteristic of a thermoelectrical generator cell, the mathematical model and the equivalent electrical model are obtained; after this step it was possible to design the TEG pack. From this point the generator is considered a general input for the power converter. Different topologies are analyzed, and all the fundamental design steps are reported for a non-inverting buck-boost converter: as part of the design, a loss estimation analysis was carried out. The control system is described as nested control loops. Two maximum power point tracking algorithms are presented illustrating their characteristics and peculiarity. A state-space averaging method is used to obtain the converter transfer function in order to design a proper controller. Finally, simulation results and converter efficiency are shown in the last section. There are many possible advancements in this field from hardware and software point of view: development of new thermoelectric cells; use of more efficient power devices, for example, Gan FETs, in order to increase frequency and reduce passive components; and, also, development of faster and more efficient control algorithm are only some examples.

Acknowledgements

The authors want to thank M. Lasana, E. Milani, and M. Brignone from Magneti Marelli Exhaust Systems group and the Mechatronics Laboratory (LIM) of the Politecnico di Torino for the support and knowledge shared.

Author details

Dario Gandini[1], Marcello Chiaberge[1*] and Andrea Nepote[2]

1 Department of Electronics and Telecommunications (DET), Politecnico di Torino, Turin, Italy

2 Magneti Marelli, Technology Innovation, Italy

*Address all correspondence to: marcello.chiaberge@polito.it

References

[1] Yu C, Chau KT. Thermoelectric automotive waste heat energy recovery using maximum power point track. Energy Conversion and Management. 2009;**50**:1506-1512. DOI: 10.1016/j.enconman.2009.02.015

[2] Liu C, Chen P, Li K. A 500 W low-temperature thermoelectric generator: Design and experimental study. International Journal of Hydrogen Energy. 2014;**39**:15497-15505. DOI: 10.1016/j.ijhydene.2014.07.163

[3] Ji D, Romagnoli A. Modelling and design of thermoelectric generator for waste heat recovery. In: ASME Proceedings 9th Symposia: Fluid Mechanics (Fundamental Issues and Perspectives; Industrial and Environmental Applications); Multiphase Flow and Systems (Multiscale Methods; Noninvasive Measurements; Numerical Methods; Heat Transfer; Performance); Transport Phenomena (Clean Energy; Mixing; Manufacturing and Materials Processing); Turbulent Flows; V01BT22A002. 2016. DOI: 10.1115/FEDSM2016-7833

[4] Esram T, Chapman PL. Comparison of photovoltaic array maximum power point tracking techniques. IEEE Transactions on Energy Conversion. June 2007;**22**(2):439-449. DOI: 10.1109/TEC.2006.874230

[5] Nagayoshi H, Kajikawa T, Sugiyama T. Comparison of maximum power point control methods for thermoelectric power generator. In: Twenty-First International Conference on Thermoelectrics (ICT '02). 2002. DOI: 10.1109/ICT.2002.1190358

[6] Laird I, Lovatt H, Savvides N, Lu D, Agelidis VG. Comparative study of maximum power point tracking algorithms for thermoelectric generators. In: 2008 Australasian Universities Power Engineering Conference. 2008. DOI: 10.1109/ICT.2002.1190358

[7] Twaha S, Zhu J, Yan Y, Li B, Huang K. Performance analysis of thermoelectric generator using dc-dc converter with incremental conductance based maximum power point tracking. Energy for Sustainable Development. 2017;**37**:86-98. DOI: 10.1016/j.esd.2017.01.003

[8] Montecucco A, Siviter J, Knox AR. Simple, fast and accurate maximum power point tracking converter for thermoelectric generators. In: 2012 IEEE Energy Conversion Congress and Exposition (ECCE). 2012. DOI: 10.1109/ECCE.2012.6342530

[9] Lee JH, Bae H, Cho BH. Advanced incremental conductance MPPT algorithm with a variable step size. In: 2006 12th International Power Electronics and Motion Control Conference. 2006. DOI: 10.1109/EPEPEMC.2006.4778466

[10] Liu B, Duan S, Liu F, Xu P. Analysis and improvement of maximum power point tracking algorithm based on incremental conductance method for photovoltaic array. In: 2007 7th International Conference on Power Electronics and Drive Systems. 2007. DOI: 10.1109/PEDS.2007.4487768